Andreas Weber

Heat Kernel Estimates and L^p-Spectral Theory of Locally Symmetric Spaces

Heat Kernel Estimates and L^p-Spectral Theory of Locally Symmetric Spaces

von
Andreas Weber

universitätsverlag karlsruhe

Dissertation, Universität Karlsruhe (TH),
Fakultät für Mathematik, 2006

Impressum

Universitätsverlag Karlsruhe
c/o Universitätsbibliothek
Straße am Forum 2
D-76131 Karlsruhe
www.uvka.de

Universitätsverlag Karlsruhe 2007
Print on Demand

ISBN: 978-3-86644-108-0

Heat Kernel Estimates and L^p-Spectral Theory of Locally Symmetric Spaces

Zur Erlangung des akademischen Grades eines

DOKTORS DER NATURWISSENSCHAFTEN

von der Fakultät für Mathematik der
Universität Karlsruhe (TH)

genehmigte

DISSERTATITON

von

Dipl.-Math. Andreas Weber
aus Gernsbach

Tag der mündlichen Prüfung:	06. Dezember 2006
Referent:	Prof. Dr. Enrico Leuzinger
Korreferent:	HDoz. Dr. Peer C. Kunstmann

Karlsruhe
2006

Preface

During the time I was working on this dissertation many people supported me. My supervisor E. Leuzinger shared with me much of his knowledge about the geometry of locally symmetric spaces. In our regular discussions he not only made many suggestions concerning my thesis but also motivated and encouraged me when I was stuck. Moreover, he supported me in my application for the Clay Summer School 2005 on "Ricci Flow, 3-Manifolds and Geometry" at the MSRI in Berkeley, USA. In the four weeks of this summer school it was very exiting to get some insight into the Ricci flow and G. Perelman's articles in which he claims to have proven Thurston's geometrization conjecture and thus, the Poincaré conjecture.
P. Kunstmann helped me a lot with the analytical part of this thesis. Whenever I dropped into his office with a question he took some time for discussion.
My colleagues from the Institute for Algebra and Geometry made the last three years a pleasant time. From my colleagues I especially need to mention G. Link who went through the whole manuscript and made many valuable comments.
E. B. Davies, W. Müller, and L. Ji answered several inquiries by e-mail.
M. E. Taylor brought his paper [75] to my knowledge.
Of course, I need also to mention my family, my friends, and last but not least Silke who supported me all the time.

It is a pleasure for me to thank all of you for your help.

Karlsruhe,
December 2006

Andreas Weber

Contents

Chapter 0.
Introduction

Why should a Riemannian geometer be interested in the study of the Laplacian? We shortly try to answer this question. Firstly, the Laplacian $\Delta := -\operatorname{div}\operatorname{grad}$ is a *canonical* elliptic differential operator of second order, which is easy to define on any Riemannian manifold. Secondly, the Laplacian, on a Riemannian manifold it is also called *Laplace-Beltrami operator*, appears in such important equations as the *heat equation* $\frac{\partial}{\partial t}u = -\Delta u$, the *wave equation* $\frac{\partial^2}{\partial t^2}u = -\Delta u$, as well as the *Schrödinger equation* $i\hbar\frac{\partial}{\partial t}u = \frac{\hbar^2}{2m}\Delta u$ of a free particle with mass m.

This dissertation is devoted to the L^p-spectral theory of the Laplace-Beltrami operator on non-compact locally symmetric spaces whose curvature is non-positive.

0.1. Heat Diffusion and Heat Equation

In chemistry or physics one type of transport phenomena is called *diffusion*. It describes the movement of particles or a substance such as heat due to concentration differences and creates a flow from regions of higher concentration to regions of lower concentration. The process of diffusion in the simplest case is modeled by the heat equation, which is a parabolic evolution equation we are now going to derive.

In 1822 Jean Baptiste Joseph Fourier stated in [34] his famous *law of heat conduction*:

$$q = -k\,\mathrm{grad} u,$$

where q is the *rate of heat flow*, $k > 0$ is called *conductivity* and $u(t, x)$ is the temperature at time t in the point x. This law forms the basis for the derivation of the heat equation (cf. for example [13]): We consider a region Ω in some perfectly isolated material and assume that there are neither heat sources nor heat sinks. Therefore, the total change of internal energy Q (in the absence of work) in Ω results from the heat flow across the boundary $\partial\Omega$. The change of internal energy Q can be related to the change of the temperature u, i.e. $Q(t_1)-Q(t_2) = c\rho(u(t_1)-u(t_2))$ with the *capacity* c and the *mass density* ρ. After the choice of a certain zero-point on the temperature scale we obtain the formula

$$Q = c\rho u.$$

Hence, the total change of internal energy in Ω is

$$c\rho \int_\Omega \frac{\partial}{\partial t} u \, dx.$$

On the other hand, using Fourier's law and the divergence theorem, the heat flow across $\partial\Omega$ into Ω is given by

$$k \int_{\partial\Omega} \langle \mathrm{grad} u, \nu \rangle \, dA = k \int_\Omega -\Delta u \, dx,$$

with an outward unit normal vector field ν on $\partial\Omega$ and the Laplacian $\Delta = -\mathrm{div}\,\mathrm{grad}$. This leads first to the equation $c\rho \frac{\partial}{\partial t} u = -k\Delta u$ and after a change of the time scale to the so called *heat equation*:

$$\frac{\partial}{\partial t} u = -\Delta u.$$

The fundamental solution $K(t,x,y)$ to this equation is called *heat kernel*, i.e. $K(t,\cdot,y)$ is a solution of the heat equation with $\lim_{t\to 0} K(t,\cdot,y) = \delta_y$, where δ_y denotes Dirac's delta (see also Chapter 2). Physically, $K(t,x,y)$ can be interpreted as the temperature in the point y at time t if the initial temperature distribution was concentrated in x with total temperature 1.

0.2. L^p-Spectral Theory

The Laplace-Beltrami operator Δ_M on a Riemannian manifold M is a positive self-adjoint operator on the Hilbert space $L^2(M)$ and $-\Delta_M$ generates therefore an analytic semigroup $e^{-t\Delta_M}$ on $L^2(M)$. Hence, for any initial heat disribution $u(0,x) = u_0(x)$ on M with $u_0 \in L^2(M)$ a solution to the heat equation is given by $u(t,x) = e^{-t\Delta_M} u_0(x)$. It can be shown (cf. Section 2) that for any $p \in [1,\infty)$ there is a strongly continuous semigroup $T_p(t)$ on $L^p(M)$ with $T_2(t) = e^{-t\Delta_M}$ such that $T_p(t)|_{L^p\cap L^q} = T_q(t)|_{L^p \cap L^q}$, i.e. the semigroups $T_p(t)$ are consistent.
On the one hand, the natural space to describe (heat) diffusion is $L^1(M)$. If

$$T_1(t)u_0 = u(t,\cdot) : M \to \mathbb{R}_{\geq 0}$$

denotes the heat distribution at time $t \geq 0$ with respect to some initial heat distribution $u_0 \in L^1(M)$, the total amount of heat in some region $\Omega \subset M$ is given by

$$||u(t,\cdot)|_\Omega||_{L^1(\Omega)} = \int_\Omega u(t,x) \, dvol(x),$$

i.e. the L^1-norm has a physical meaning.
On the other hand, the L^1-spaces turn out to be more difficult to handle than the reflexive L^p-spaces. One example for this is the question whether the semigroup $T_1(t)$ is analytic or not, whereas for $p \in (1, \infty)$ the semigroups $T_p(t)$ are always analytic.
We denote by $-\Delta_{M,p}$ the generator of the semigroup $T_p(t)$ and call $\Delta_{M,p}$ the Laplace-Beltrami operator on $L^p(M)$. It is actually possible that the L^p-spectrum $\sigma(\Delta_{M,p})$ depends on p. This may happen because, in spite of the consistency of the semigroups $T_p(t)$, the resolvents $(\lambda - \Delta_{M,p})^{-1}$ need not be consistent for some λ.

For the L^p-spectral theory of more general (elliptic) operators on $L^p(\Omega)$, where Ω denotes an open subset of $\mathbb{R}^n$, we refer to [3, 21, 43, 44, 46, 52, 53, 54].

0.2.1. Volume Growth and L^p-Spectrum

K.-T. Sturm related in [72] the question whether the spectrum $\sigma(\Delta_{M,p})$ depends on p or not to the volume growth of the Riemannian manifold M. He actually considered not only the Laplace-Beltrami operator on M but also uniformly elliptic operators in divergence form. In terms of the Laplace-Beltrami operator $\Delta_{M,p}$ his main results can be stated as follows.

Theorem 0.1. (p-Independence). *Let M denote a Riemannian manifold with Ricci curvature bounded from below. If the volume of M grows* uniformly subexponentially, *i.e. for any $\varepsilon > 0$ there is some constant $C > 0$ such that for all $x \in M$ and $r > 0$ the following inequality for the volume of balls $B(x, r)$ with center x and radius r holds:*

$$\operatorname{vol} B(x, r) \leq C e^{\varepsilon r} \operatorname{vol} B(x, 1),$$

then we have for $p \in [1, \infty)$:

$$\sigma(\Delta_{M,p}) = \sigma(\Delta_{M,2}).$$

We conclude that the L^p-spectrum of a Riemannian manifold M does not depend on p if M is compact or if the Ricci curvature of M is non-negative. In particular, the L^p-spectrum of the Laplacian on Euclidean space $\mathbb{R}^n$ does not depend on p (and coincides with the interval $[0, \infty)$).

To state the next theorem we need the following definition. Let us denote by $S_x M$ the unit sphere in the tangent space $T_x M$ of M in x and by $\sqrt{g(r, \zeta)}$, $r \geq 0$ and $\zeta \in S_x M$, the volume density of the Riemannian manifold M with respect to geodesic normal coordinates. We say that the *volume density of M grows exponentially in every direction* if there is some point $x \in M$ with empty cut locus and constants $\varepsilon, C > 0$ such that

$$\sqrt{g(r, \zeta)} \geq C e^{\varepsilon r},$$

for any $r > 0$ and $\zeta \in S_x M$.

Theorem 0.2. (p-Dependence). *Let M denote a Riemannian manifold with Ricci curvature bounded from below. If the volume density of M grows* exponentially *in every direction, it follows*

$$\inf \operatorname{Re} \sigma(\Delta_{M,1}) = 0 \qquad \text{and} \qquad \inf \sigma(\Delta_{M,2}) > 0.$$

In particular, the L^p-spectrum depends on p.

Notice, that all simply connected manifolds with constant negative curvature satisfy the conditions in Theorem 0.2. In particular, the L^p-spectrum of the Laplace-Beltrami operator on the n-dimensional hyperbolic space $\mathbb{H}^n$ ($n \geq 2$) depends on p. However, it is not at all clear how the L^p-spectrum precisely looks like. Another point to mention is that Theorem 0.2 can not be applied to non-trivial quotient manifolds, say e.g. $\Gamma\backslash\mathbb{H}^n$ where $\Gamma \subset \operatorname{Isom}(\mathbb{H}^n)$ denotes a discrete subgroup of the isometry group which acts freely on $\mathbb{H}^n$, since there is no point in $\Gamma\backslash\mathbb{H}^n$ with empty cut locus.

In the case of non-compact quotients $M = \Gamma\backslash\mathbb{H}^n$ where Γ denotes a geometrically finite subgroup, E. B. Davies, B. Simon, and M. E. Taylor explicitly determined the L^p-spectrum of the respective Laplace-Beltrami operator ([24]). They prove that the L^p-spectrum $\sigma(\Delta_{M,p})$ is the union of a (possibly empty) discrete set of eigenvalues of $\Delta_{M,2}$ and a parabolic region P_p that degenerates to the interval $[(n-1)^2/4, \infty)$ for $p = 2$:

$$\sigma(\Delta_{M,p}) = \{\lambda_0, \ldots, \lambda_m\} \cup P_p.$$

For quotients $M = \Gamma\backslash\mathbb{H}^n$ with infinite volume the authors needed also to assume that M has no cusps.

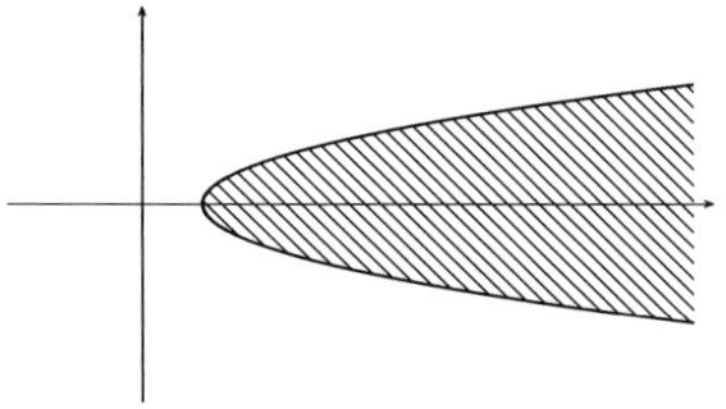

Figure 0.1.: L^p-Spectrum of $\mathbb{H}^n$.

M. E. Taylor generalized the mentioned results from [24] to symmetric spaces of non-compact type (cf. [75]). He also showed that the methods from [24] can be used to prove that the L^p-spectrum of certain locally symmetric spaces M is contained in the union of some discrete set of eigenvalues for $\Delta_{M,2}$ and some parabolic region.

We conclude this section with a result due to E. B. Davies (cf. [20]), that in some points could be regarded as a generalization of Theorem 0.1. Davies considers, more generally, a locally compact, second countable metric space M but assumes the somewhat stronger condition of *polynomial volume growth* instead of uniform subexponential volume growth. If A denotes a non-negative self-adjoint operator on $L^2(M)$ (defined with respect to some Borel measure on M) such that the semigroup $e^{-tA} : L^2(M) \to L^2(M)$ has an integral kernel with a Gaussian upper bound, he shows that there is a consistent family of strongly continuous semigroups $T_p(t)$ on $L^p(M)$, $p \in [1, \infty)$, with $T_2(t) = e^{-tA}$. Furthermore, if $-A_p$ denotes the generator of $T_p(t)$, the spectrum $\sigma(A_p)$ does *not* depend on p.

0.3. Locally Symmetric Spaces

Why do we restrict ourselves to the Laplace-Beltrami operator on locally symmetric spaces with non-positive curvature?

A substantial part of this dissertation was inspired by the already cited paper [24]. I was wondering to what extent these results could be generalized to manifolds with variable curvature. In order to obtain precise results the manifolds in our considerations should not be too general.

Locally symmetric spaces are characterized by $\nabla R = 0$, i.e. the covariant derivative of the Riemannian curvature tensor R vanishes or, equivalently, the sectional curvature is invariant under parallel translations (cf. Chapter 8, Proposition 10 in [63]).

Hence, the class of locally symmetric spaces includes all manifolds with constant curvature and in particular quotients $\Gamma\backslash\mathbb{H}^n$ of the n-dimensional hyperbolic space by a discrete subgroup Γ of the isometry group, that acts freely on $\mathbb{H}^n$, which were considered in [24]. On the other hand, locally symmetric spaces have enough structure in order to make precise statements. Actually, a complete classification of (globally) symmetric spaces due to Élie Cartan is available (see e.g. [41]).

Note also, that the L^2-spectral theory of locally symmetric spaces plays a major role in fields like harmonic analysis, representation theory, and number theory.

0.4. Outline of Chapter 1 – Chapter 6

The first two chapters have an introductory character. Our main results are contained in the Chapters 3 – 5.

Chapter 1.

We introduce locally symmetric spaces and present some decomposition theorems needed in the following.

Chapter 2.

Basic properties of the heat kernel on Riemannian manifolds are stated in the first section. We define the heat semigroup on L^p-spaces for $p \in [1, \infty)$ as well as the Laplace-Beltrami operator on L^p. In Section 2.3 we prove that the L^p-spectrum of a Riemannian product $M_1 \times M_2$ is nothing else than the set theoretic sum of the L^p-spectra of M_1 and M_2, $p \in (1, \infty)$. This result enables us to restrict ourselves to irreducible Riemannian manifolds if we are interested in the L^p-spectrum. Optimal bounds for the heat kernel on globally symmetric spaces of non-positive curvature are discussed in Section 2.4. These results, in their final form, are due to J.-P. Anker and P. Ostellari (cf. [2, 64]). In the last section of Chapter 2 we prove a formula connecting the heat kernel K of a Cartan-Hadamard manifold X with the heat kernel k of a quotient $M = \Gamma\backslash X$ of X by a discrete subgroup $\Gamma \subset \mathrm{Isom}(X)$ of the isometry group that acts freely on X. More precisely, we show for all $x, y \in X$:

$$k(t, \pi(x), \pi(y)) = \sum_{\gamma\in\Gamma} K(t, x, \gamma y), \tag{0.1}$$

where $\pi : X \to \Gamma\backslash X$ denotes the covering map. In case of $X = \mathbb{R}$ and $M = S^1 = \mathbb{R}/\mathbb{Z}$ formula (0.1) reads

$$k(t, \mathrm{e}^{ix}, \mathrm{e}^{iy}) = \sum_{n\in\mathbb{Z}} K(t, x, y + 2\pi n),$$

where $x, y \in \mathbb{R}$. Physically, we expect this formula to be true as heat starting in $\mathrm{e}^{ix} \in S^1$ can arrive at $\mathrm{e}^{iy} \in S^1$ by flowing around the circle in any direction any number of times.

Chapter 3.

In this chapter we derive (Gaussian) upper bounds for the heat kernel on locally symmetric spaces $M = \Gamma\backslash X$ and lower bounds for the bottom $\lambda_0(M)$ of the L^2-spectrum $\sigma(\Delta_{M,2})$. Our main results are stated in Theorem 3.15, Corollary 3.22, and Corollary 3.24. This generalizes results from [23] and gives a new proof for Leuzinger's lower bounds of $\lambda_0(M)$ (see [57]).
We begin with a brief discussion of the Laplace-Beltrami operator on a *general* compact Riemannian manifold. Since such general statements as in the first section can not be expected in the non-compact case we restrict ourselves to non-compact locally symmetric spaces in the remaining part.
A major role in the derived upper bounds for the heat kernel plays the so-called *Poincaré series*

$$P(s; x, y) := \sum_{\gamma\in\Gamma} \mathrm{e}^{-sd(x,\gamma y)},$$

with $s \in (0, \infty)$ and $x, y \in X$. Therefore, we are concerned with estimates of the Poincaré series in Section 3.2. In Section 3.3 we derive upper bounds for the heat kernel k on $\Gamma\backslash X$ by estimating each term on the right hand side of formula (0.1) using Anker's and Ostellari's upper bound for the heat kernel K on X (Theorem 2.12). In the upper bounds of $k(t, \pi(x), \pi(y))$ obtained in this way appears the Poincaré series $P(s; x, y)$ (see e.g. Theorem 3.6). Unfortunately, our estimates from Section 3.2 do only apply to Poincaré series of the form $P(s; x, x)$. On the other hand, the results can be used to derive lower bounds for $\lambda_0(M)$ (Section 3.4), and these bounds in turn, together with a result due to Davies and Mandouvalos (Theorem 3.19), yield new heat kernel bounds (Corollary 3.22 and Corollary 3.24) containing the functions $P(s; x, x)$ instead of $P(s; x, y)$.

Chapter 4.

Our main result in this chapter is Theorem 4.2, in which we completely determine the L^p-spectrum of the Laplace-Beltrami operator on certain locally symmetric spaces $M = \Gamma\backslash X$. More precisely, our theorem states that $\sigma(\Delta_{M,p})$ coincides with a parabolic region P_p depending on p:

Theorem 4.2. *Let $X = G/K$ denote a symmetric space of non-compact type with* $\mathrm{rank}(X) = 1$ *and* $\dim(X) \geq 3$. *Assume, that the locally symmetric space $M = \Gamma\backslash X$ has bounded geometry and that Γ is small. Then we have for $p \in [1, \infty)$:*

$$\sigma(\Delta_{M,p}) = P_p = \sigma(\Delta_{X,p}).$$

In the proof of this theorem we use the heat kernel estimates and the lower bounds for $\lambda_0(M)$ derived in Chapter 3.
If $p = 2$, the statement of Theorem 4.2 reads as follows:

$$\sigma(\Delta_{M,2}) = [||\rho||^2, \infty) = \sigma(\Delta_{X,2}).$$

For a definition of ρ see Chapter 1. Even in this case the result seems to be new. Theorem 4.2 is a *rigidity theorem* for the L^p-spectra of locally symmetric spaces $\Gamma\backslash X$ with respect to *small* subgroups Γ, where the size of a discrete subgoup Γ is measured by the critical exponent $\delta(\Gamma)$ of the Poincaré series (cf. Section 3.2): If Γ is as in Theorem 4.2, the L^p-spectrum $\sigma(\Delta_{M,p})$ of $M = \Gamma\backslash X$ coincides always with the L^p-spectrum of the universal cover X of M.
Note, that the locally symmetric spaces from above all have infinite volume (Theorem 4.19). A class of non-compact locally symmetric spaces with finite volume will be treated in Chapter 5.

Chapter 5.

We examine the L^p-spectrum of a locally symmetric space $M = \Gamma\backslash X$ where Γ denotes an arithmetic lattice with $\mathbb{Q}$-rank 1 that acts freely on a symmetric space X of non-compact type.

If X is a rank one symmetric space, we again are able to determine the L^p-spectrum of the Laplace-Beltrami operator $\Delta_{M,p}$. In this case we have for $p \in (1, \infty)$

$$\sigma(\Delta_{M,p}) = \{\lambda_0, \dots, \lambda_m\} \cup P_p,$$

where $0 = \lambda_0, \dots, \lambda_m \in [0, ||\rho||^2)$ are eigenvalues of $\Delta_{M,2}$ and P_p is the same parabolic region as in Chapter 4, i.e. $P_p = \sigma(\Delta_{X,p})$. This result is stated in Corollary 5.15. Consequently, the continuous spectrum of the Laplace-Beltrami operator on $L^p(\Gamma\backslash X)$ is, for fixed X with rank one, for all considered non-compact locally symmetric spaces $\Gamma\backslash X$ the same.
In the case where X denotes a higher rank symmetric space we are able to prove that a certain parabolic region is contained in the L^p-spectrum. But now, the parabolic region coincides in general *not* with $\sigma(\Delta_{X,p})$. Nevertheless, it seems to be likely that the L^p-spectrum of M in this case is the union of some finite set $\{\lambda_0, \dots, \lambda_m\}$ and this (different) parabolic region.

Chapter 6.

The proofs in Chapter 5 show that the cusps of a locally symmetric space $\Gamma\backslash X$ are responsible for the parabolic region contained in the L^p-spectrum. Hence, the results of the preceding chapter generalize immediately to manifolds with cusps of rank one. This generalization is briefly discussed in this chapter.

Chapter 1.

Locally Symmetric Spaces

A connected Riemannian manifold X is called *symmetric space* if for all $x \in X$ there is an isometry $s_x \in \mathrm{Isom}(X)$ such that

(i) $s_x(x) = x$,

(ii) $ds_x|_x = -\mathrm{id}_{T_xX}$.

The isometry s_x is called *geodesic symmetry* at the point $x \in X$. Since isometries preserve geodesics, we obtain immediately that for any geodesic $c(t)$ with $c(0) = x$ the formula $s_x(c(t)) = c(-t)$ holds. As a first consequence symmetric spaces are always complete and homogeneous. Furthermore, we have for all $x \in X$:

$$s_x : X \to X, \ \exp_x(v) \mapsto \exp_x(-v),$$

where $\exp_x : T_xX \to X$ denotes the Riemannian exponential map. In the following, we will only consider simply connected symmetric spaces X with non-positive sectional curvature. Then we have the decomposition $X = X_- \times X_0$, where X_- denotes a symmetric space of non-compact type and X_0 a Euclidean space (cf. Proposition 4.2 in [41]).

1.1. Algebraic Description of Symmetric Spaces

As symmetric spaces are homogeneous, there is a description as coset manifolds $X = G/K$ where G and K are certain Lie groups with K being compact in G. In the subsequent sections we will need more information about the structure of the Lie group G, and therefore we summarize several important properties:
Since X is a product $X_- \times X_0$, the identity component $\mathrm{Isom}^0(X)$ of the isometry group $\mathrm{Isom}(X)$ also splits (cf. Theorem VI.3.5 in [51]):

$$\mathrm{Isom}^0(X) = \mathrm{Isom}^0(X_-) \times \mathrm{Isom}^0(X_0).$$

Then $G := \mathrm{Isom}^0(X_-) \times \mathbb{R}^m \subset \mathrm{Isom}(X)$ is a connected real reductive Lie group in the Harish-Chandra class and still acts transitively on our symmetric space X. Recall, that the attribute *Harish-Chandra class* means the following: If we denote by $\mathfrak{g}$ the Lie algebra of G then the semi-simple Lie group with Lie algebra $[\mathfrak{g}, \mathfrak{g}]$ has finite center.
Furthermore, the isotropy subgroup $K := \{g \in G : g \cdot x_0 = x_0\}$ of an arbitrary basepoint x_0 is a maximal compact subgroup of G and X is diffeomorphic to the quotient G/K.

If we consider on the other hand a connected real reductive non-compact Lie group G in the Harish-Chandra class and a maximal compact subgroup $K \subset G$, the coset manifold G/K admits a G-invariant Riemannian metric relative to which G/K is a symmetric space with non-positive curvature (this is actually true for any G-invariant metric). In the following, we give a short outline of the proof of this statement to point out the differences from the semi-simple setting. Our main references are Part II of [76] and [35, 49].
With regard to Proposition IV.3.4 in [41] it is sufficient to find, besides a G-invariant metric, an analytic involution σ of G such that the set of fixed points of the involution σ coincides with K. But this is the content of Theorem 13 in Part II of [76]. This involution induces further a *Cartan involution* $\theta := d\sigma$ of the Lie algebra $\mathfrak{g}$ of G and therefore a decomposition of $\mathfrak{g}$ – the *Cartan decomposition* – into the respective eigenspaces of θ:

Theorem 1.1. *Let $\mathfrak{g}$ denote the Lie algebra of a real reductive Lie group G in the Harish-Chandra class. We have*

$$\mathfrak{g} = \mathfrak{k} \oplus \mathfrak{p},$$

where $\mathfrak{k} := \{Z \in \mathfrak{g} : \theta Z = Z\}$ and $\mathfrak{p} := \{Z \in \mathfrak{g} : \theta Z = -Z\}$. Furthermore, $\mathfrak{k}$ is the Lie algebra of the maximal compact subgroup K.

Proof. The decomposition follows immediately from $\theta^2 = \mathrm{id}$ and the assertion that the Lie algebra of K is given by $\mathfrak{k}$ is analogously proven as Theorem IV.3.3 in [41]. □

The fact that θ is a Lie algebra automorphism implies

$$[\mathfrak{k}, \mathfrak{k}] \subset \mathfrak{k}, \quad [\mathfrak{k}, \mathfrak{p}] \subset \mathfrak{p}, \quad [\mathfrak{p}, \mathfrak{p}] \subset \mathfrak{k}. \tag{1.1}$$

Since the Lie algebra $\mathfrak{g}$ is reductive, we also have the decomposition

$$\mathfrak{g} = \mathfrak{g}_1 \oplus Z(\mathfrak{g}),$$

where $\mathfrak{g}_1 := [\mathfrak{g}, \mathfrak{g}]$ is a semi-simple Lie algebra and $Z(\mathfrak{g}) := \{H \in \mathfrak{g} : [H, Y] = 0 \text{ for all } Y \in \mathfrak{g}\}$ denotes the center of $\mathfrak{g}$. The Lie subgroup G_1 of G with Lie algebra

$\mathfrak{g}_1$ has finite center because G lies in the Harish-Chandra class. The intersection $K_1 := G_1 \cap K$ is a maximal compact subgroup in G_1. This induces a Cartan decomposition $\mathfrak{g}_1 = \mathfrak{k}_1 \oplus \mathfrak{p}_1$ of the semi-simple Lie algebra $\mathfrak{g}_1$. Because of $Z(\mathfrak{g}) = (\mathfrak{k} \cap Z(\mathfrak{g})) \oplus (\mathfrak{p} \cap Z(\mathfrak{g}))$ (use (1.1)), we can conclude

$$\begin{aligned} \mathfrak{k} &= \mathfrak{k}_1 \oplus (\mathfrak{k} \cap Z(\mathfrak{g})), \\ \mathfrak{p} &= \mathfrak{p}_1 \oplus (\mathfrak{p} \cap Z(\mathfrak{g})). \end{aligned}$$

The definition of a G-invariant metric on G/K is equivalent to the definition of an $\mathrm{Ad}(K)$-invariant inner product on $\mathfrak{p} \cong T_{eK}(G/K)$. The next theorem ensures the existence of such an inner product.

Theorem 1.2 (cf. Theorem 16 in Part II of [76]). *Let $\mathfrak{g}$ denote the Lie algebra of a real reductive Lie group G in the Harish-Chandra class. Then there exists a non-degenerate symmetric bilinear form B on $\mathfrak{g}$ such that*

(i) *B is invariant under $\mathrm{Ad}(G)$ and θ,*

(ii) *B is negative definite on $\mathfrak{k}$ and positive definite on $\mathfrak{p}$,*

(iii) $B(\mathfrak{k}, \mathfrak{p}) = B(\mathfrak{g}_1, Z(\mathfrak{g})) = \{0\}$.

In particular, if we define for Y and $Z \in \mathfrak{g}$

$$\langle Y, Z \rangle := -B(Y, \theta Z),$$

then $\langle \cdot, \cdot \rangle$ is an $\mathrm{Ad}(K)$-invariant inner product on $\mathfrak{g}$.

Proof. We define

$$B := B_1 \oplus B_2,$$

where B_1 denotes the Killing form on $\mathfrak{g}_1$ and B_2 an arbitrary symmetric non-degenerate bilinear form on $Z(\mathfrak{g})$ which is negative definite on $\mathfrak{k} \cap Z(\mathfrak{g})$, positive definite on $\mathfrak{p} \cap Z(\mathfrak{g})$, and for which the subspaces $\mathfrak{k} \cap Z(\mathfrak{g})$ and $\mathfrak{p} \cap Z(\mathfrak{g})$ are orthogonal. Then (ii), (iii) and the θ-invariance are evident. To show the $\mathrm{Ad}(G)$-invariance of B, we first claim that $\mathrm{Ad}(G)$ acts trivially on $Z(\mathfrak{g})$. More precisely, we want to show for all $Y \in Z(\mathfrak{g})$ and $g \in G$ the identity

$$\mathrm{Ad}(g)Y = Y.$$

To see this, we first mention that for $Y \in Z(\mathfrak{g})$ the image $\exp(tY)$ lies in the center $Z(G)$ of G: One checks easily $\exp(tY)g = g\exp(tY)$ for $g = \exp(Y') \in U$ where U is a neighborhood of e in G such that the exponential map $\exp : V \to U$ is a diffeomorphism for a certain neighborhood V of 0 in $\mathfrak{g}$. Since U generates the connected group G the same is true for arbitrary $g \in G$. The claim now follows from

$$\exp(t\mathrm{Ad}(g)Y) = g\exp(tY)g^{-1} = \exp(tY).$$

To complete the proof we choose open neighborhoods $V := V_1 \oplus V_2$ of 0 in $\mathfrak{g} = \mathfrak{g}_1 \oplus Z(\mathfrak{g})$ and U of e in G such that the map $\varphi : V := V_1 \oplus V_2 \to U, (Y_1, Y_2) \mapsto \exp(Y_1)\exp(Y_2)$ is a diffeomorphism. For $g \in U$ we therefore have a unique decomposition $g = g_1 g_2$ with $g_1 \in G_1$ and $g_2 \in Z(G)$. For all $Y \in \mathfrak{g}_1$ we can conclude

$$\mathrm{Ad}(g)Y = \mathrm{Ad}(g_1)Y,$$

where the equality holds because the center $Z(G)$ of G equals the kernel of the adjoint representation $\mathrm{Ad} : G \to GL(\mathfrak{g})$. Now the $\mathrm{Ad}(G)$-invariance of B follows from the $\mathrm{Ad}(G_1)$-invariance of the Killing form B_1 of the semi-simple Lie group G_1. □

In conclusion we obtain a decomposition of our symmetric space $X := G/K$ as a product of the symmetric space $X_1 := G_1/K_1$ of non-compact type and the Euclidean space $X_2 \cong (\mathfrak{p} \cap Z(\mathfrak{g}))$.

1.2. The Metric of Irreducible Symmetric Spaces

In the preceding theorem we constructed – by using the Killing form – a Riemannian metric that turns G/K into a symmetric space of non-positive curvature. Now it is natural to ask how many other metrics possibly exist. Because of de Rahm's decomposition theorem it suffices to answer this question for *irreducible symmetric spaces G/K of non-compact type* only. Notice that in this case G is a simple Lie group. Then we have the following result.

Theorem 1.3. *Assume G is a non-compact simple Lie group, $K \subset G$ a maximal compact subgroup and g a Riemannian metric that turns G/K into a symmetric space of non-compact type. Then the restriction of g to $\mathfrak{p} \cong T_{eK}(G/K)$ is a (positive) multiple of the Killing form B on $\mathfrak{g}$, i.e.*

$$g_{eK}(Y, Z) = \lambda B(Y, Z)$$

for some $\lambda > 0$ and all $Y, Z \in \mathfrak{p}$.

Proof. The assertion is equivalent to the analogous statement about orthogonal involutive simple Lie algebras. The proof therefore follows directly from Theorem 8.2.9 in [80]. □

1.3. Locally Symmetric Spaces

A Riemannian manifold M is called a *locally symmetric space* if for any point $p \in M$ there exists a neighborhood U of p such that the local geodesic symmetry s_p at p, defined by

$$s_p : U \to U, \exp_p(v) \to \exp_p(-v),$$

is an isometry.

If G denotes a non-compact real reductive Lie group in Harish-Chandra's class, $K \subset G$ a maximal compact subgroup, and $\Gamma \subset G$ a discrete subgroup of isometries which acts freely on $X = G/K$, the orbit space $M := \Gamma\backslash X$ is a complete locally symmetric space. The canonical projection $\pi : X \to M$, $x \mapsto \Gamma x$ is a Riemannian covering and Γ is isomorphic to the fundamental group $\pi_1(M)$ of M. The metric d_M on M is given by $d_M(\Gamma x, \Gamma y) = \min_{\gamma\in\Gamma} d(x, \gamma y)$ and is just the distance of the respective orbits in X.

1.4. Decomposition Theorems

Let $\mathfrak{g}$ denote the Lie algebra of a real reductive Lie group G in the Harish-Chandra class. We already became acquainted with the *Cartan decomposition*

$$\mathfrak{g} = \mathfrak{k} \oplus \mathfrak{p}$$

of the Lie algebra $\mathfrak{g}$ into eigenspaces with respect to a *Cartan involution* θ of $\mathfrak{g}$. Below we will need other decompositions of both the reductive Lie algebra $\mathfrak{g}$ and the reductive Lie group G. Our main references are again [35], Part II of [76] or [49].

Let us fix a Cartan decomposition $\mathfrak{g} = \mathfrak{k} \oplus \mathfrak{p}$ and choose a maximal abelian subalgebra $\mathfrak{a} \subset \mathfrak{p}$. Recall that all such maximal abelian subalgebras are conjugate under K. The dimension of such an algebra does therefore not depend on our choice and we define the rank of the symmetric space G/K by

$$\operatorname{rank}(G/K) := \dim \mathfrak{a}.$$

We further denote by $\mathfrak{a}^*$ the dual space of $\mathfrak{a}$ and put for any $\alpha \in \mathfrak{a}^*$

$$\mathfrak{g}_\alpha := \{Y \in \mathfrak{g} : \operatorname{ad}(H)(Y) = \alpha(H)Y \text{ for all } H \in \mathfrak{a}\}.$$

Then $\alpha \in \mathfrak{a}^* \setminus \{0\}$ is called a *(restricted) root* of $(\mathfrak{g}, \mathfrak{a})$ if $\mathfrak{g}_\alpha \neq \{0\}$. Let us denote by $\Sigma = \Sigma(\mathfrak{g}, \mathfrak{a})$ the set of all (restricted) roots. If α is a root, the only multiples of α that can also be roots are $\pm\frac{1}{2}\alpha, \pm\alpha, \pm 2\alpha$, and $-\alpha$ is always a root with $\dim \mathfrak{g}_\alpha = \dim \mathfrak{g}_{-\alpha}$. In contrast to the semi-simple case there can be a non-trivial subspace of $\mathfrak{a}$ on which all the roots vanish: we have $\alpha(H) = 0$ for all $\alpha \in \Sigma$ if and only if $H \in Z(\mathfrak{g}) \cap \mathfrak{p}$.

One easily checks that the family $\{\operatorname{ad}(H) : H \in \mathfrak{a}\}$ consists of commuting and symmetric operators with respect to the non-degenerate bilinear form $\langle\cdot,\cdot\rangle$ defined in Theorem 1.2. As a consequence of this we obtain the following.

Theorem 1.4. (Root space decomposition). *The reductive Lie algebra* $\mathfrak{g}$ *is the direct sum of* root spaces*:*

$$\mathfrak{g} = \mathfrak{g}_0 \oplus \bigoplus_{\alpha\in\Sigma} \mathfrak{g}_\alpha.$$

Notice that we have $\mathfrak{g}_0 = \mathfrak{m} \oplus \mathfrak{a}$ where $\mathfrak{m} := Z_{\mathfrak{g}}(\mathfrak{a}) \cap \mathfrak{k}$ is the intersection of the centralizer $Z_{\mathfrak{g}}(\mathfrak{a})$ of $\mathfrak{a}$ in $\mathfrak{g}$ with the Lie subalgebra $\mathfrak{k}$.

We call an $H \in \mathfrak{a}$ *regular* if $\alpha(H) \neq 0$ for all $\alpha \in \Sigma$, otherwise *singular*. The subset of regular elements

$$\mathfrak{a}_{reg} := \{H \in \mathfrak{a} \ : \ \alpha(H) \neq 0 \text{ for all } \alpha \in \Sigma\}$$

is the complement of a union of finitely many hyperplanes and the connected components of $\mathfrak{a}_{reg}$ are called *Weyl chambers*. Let us fix a Weyl chamber $\mathfrak{a}^+$. With respect to this Weyl chamber a root α is said to be positive if $\alpha(H) > 0$ for all $H \in \mathfrak{a}^+$. We denote by Σ^+ the *set of positive roots* and by Σ_0^+ the *set of indivisible positive roots*, where a positive root α is called indivisible if $\frac{1}{2}\alpha$ is not a root. Then

$$\mathfrak{n} := \bigoplus_{\alpha \in \Sigma^+} \mathfrak{g}_\alpha$$

is a nilpotent subalgebra of $\mathfrak{g}$. We further denote by $N := \exp \mathfrak{n}$ the analytic subgroup of G defined by $\mathfrak{n}$.

Theorem 1.5. (Iwasawa decomposition *of* $\mathfrak{g}$ *and* G**).**

(a) *We have the following direct sum decompostion of* $\mathfrak{g}$*:*

$$\mathfrak{g} = \mathfrak{k} \oplus \mathfrak{a} \oplus \mathfrak{n}.$$

(b) *The map*

$$K \times A \times N \to G, \ (k, a, n) \mapsto kan$$

is a diffeomorphism.

We will write $\log : A \to \mathfrak{a}$ for the inverse of the diffeomorphism $\exp : \mathfrak{a} \to A$. Notice that for a given K the Iwasawa decomposition depends on the choice of a maximal abelian subalgebra $\mathfrak{a}$ in $\mathfrak{p}$ and on the choice of a Weyl chamber $\mathfrak{a}^+$ in $\mathfrak{a}$.

Let us denote by $m_\alpha := \dim \mathfrak{g}_\alpha$ the multiplicity of a root α. We further define

$$\rho := \frac{1}{2} \sum_{\alpha \in \Sigma^+} m_\alpha \alpha$$

as half the sum of the positive roots counted according to their multiplicity.

The Cartan decomposition $\mathfrak{g} = \mathfrak{k} \oplus \mathfrak{p}$ induces a decomposition $G = KP$ with $P := \exp \mathfrak{p}$. Since all maximal abelian subalgebras $\mathfrak{a} \subset \mathfrak{p}$ are conjugate under K, we clearly have $\mathfrak{p} = \bigcup_{k \in K} \mathrm{Ad}(k)\mathfrak{a}$ and therefore $P \subset KAK$. This proves the following proposition.

Proposition 1.6. *For a non-compact real reductive Lie group G in the Harish-Chandra class we have the decomposition*

$$G = KAK.$$

In the following, we want to give a refinement of this proposition. For this, we first define the *Weyl group*

$$W := M'/M$$

of the pair $(\mathfrak{g}, \mathfrak{a})$ with the normalizer $M' := \{k \in K : \mathrm{Ad}(k)\mathfrak{a} \subset \mathfrak{a}\}$ and the centralizer $M := \{k \in K : \mathrm{Ad}(k)H = H \text{ for all } H \in \mathfrak{a}\}$ of $\mathfrak{a}$ in K. It is known that W acts simply transitively on the set of Weyl chambers. Given two Weyl chambers $\mathfrak{a}_1^+$ and $\mathfrak{a}_2^+$ we can therefore find a $k \in K$ such that $\mathrm{Ad}(k)\mathfrak{a}_1^+ = \mathfrak{a}_2^+$. Together with Proposition 1.6 this proves the first part of the following Theorem. For the uniqueness we refer to Lemma 2.2.3 in [35].

Theorem 1.7. (Cartan decomposition *of* G**).** *Let G denote a non-compact real reductive Lie group in the Harish-Chandra class and $\overline{\mathfrak{a}^+}$ the closure of the Weyl chamber $\mathfrak{a}^+$. Then we have the decomposition*

$$G = K \exp \overline{\mathfrak{a}^+} K.$$

More precisely, this means that each $g \in G$ can be written as $g = k_1 \exp(H) k_2$ with $k_1, k_2 \in K$ and $H \in \overline{\mathfrak{a}^+}$. Moreover, $H = H(g)$ is unique.

Chapter 2.

Heat Kernels on Riemannian Manifolds and Spectral Theory

In this Chapter we give a short introduction to the theory of heat kernels, the Laplace-Beltrami operator on L^p-spaces, and we derive a formula for the heat kernel on quotients of Cartan-Hadamard manifolds which will be of importance in Chapter 3.

2.1. The Heat Kernel on a Riemannian Manifold

Let us denote by (M, g) an arbitrary Riemannian manifold and by

$$\Delta_M := -\operatorname{div} \operatorname{grad}$$

the corresponding Laplace-Beltrami operator on $C^2(M)$. In local coordinates we have the formula

$$\Delta_M = -\frac{1}{\sqrt{\det(g_{ij})}} \sum_{i,j} \partial_i (g^{ij} \sqrt{\det(g_{ij})}\, \partial_j), \tag{2.1}$$

where the metric g is represented by the matrix (g_{ij}) in these coordinates and the inverse matrix is denoted by (g^{ij}). The principal symbol of Δ_M is $\xi \mapsto -\sum_{i,j} g^{ij} \xi_i \xi_j$ and Δ_M is an elliptic operator (of order 2).

The *heat kernel* $K(t, x, y)$ on M is defined as the smallest positive fundamental solution to the heat equation on $(0, \infty) \times M$. This means that for a bounded and continuous function $u_0 : M \to \mathbb{R}$ a solution to the Cauchy problem

$$\begin{aligned} \frac{\partial}{\partial t} u(t, x) &= -\Delta_M u(t, x), \\ u(0, x) &= u_0(x) \end{aligned} \tag{2.2}$$

is given by

$$u(t,x) = \int_M K(t,x,y)u_0(y)\,dvol(y).$$

We say that $u : [0,T) \times M \to \mathbb{R}$ is a *solution of the Cauchy problem* (2.2) with initial data u_0, if

(i) the function u is continuous,

(ii) u is C^1 in t for $t \in (0,T)$ and C^2 in the space component,

(iii) $u(0,\cdot) = u_0$,

(iv) u satisfies the heat equation: $\frac{\partial}{\partial t}u(t,x) = -\Delta_M u(t,x)$.

There exists always a heat kernel K on a Riemannian manifold (cf. [14]). We summarize several important properties in the next lemma.

Lemma 2.1. *For the heat kernel K on a Riemannian manifold M the following holds:*

(a) *K is a strictly positive C^∞-function on $(0,\infty) \times M \times M$,*

(b) *K is symmetric in the space components,*

(c) $\int_M K(t,x,y)\,dvol(y) \leq 1$,

(d) $\int_M K(s,x,y)K(t,y,z)\,dvol(y) = K(s+t,x,z)$ *(semigroup property).*

The proof of these properties is contained in Chapter VIII of [14].

We also have the following uniqueness theorem (c.f. Theorem 2.2 in [26]):

Theorem 2.2. *If M is a complete Riemannian manifold with Ricci curvature bounded from below, then bounded solutions to the Cauchy problem (2.2) are uniquely determined by their initial data u_0.*

That some curvature condition in this theorem is necessary is a consequence of [4], Proposition 7.9: R. Azencott proved that for complete, simply connected Riemannian manifolds with negative sectional curvature which tends to $-\infty$ sufficiently fast, there exists a non-constant solution u of the Cauchy problem with initial data $u_0 = 1$.

From the uniqueness theorem we obtain immediately that there is exactly one fundamental solution to the heat equation if the manifold is complete and the Ricci curvature is bounded from below. It follows also that these manifolds are *stochastically complete*:

Lemma 2.3. *Let M denote a complete Riemannian manifold with Ricci curvature bounded from below. Then the heat kernel K satisfies*

$$\int_M K(t,x,y)\,dvol(y) = 1.$$

Recall that locally symmetric spaces are complete Riemannian manifolds with Ricci curvature bounded from below since their universal covering spaces are homogeneous. Thus, the above results apply to these spaces.

Example 2.4. In the Euclidean case $M = \mathbb{R}^n$ the heat kernel is given by the well known formula

$$K(t,x,y) = \frac{1}{(4\pi t)^{n/2}} \exp\left(-\frac{d^2(x,y)}{4t}\right).$$

If $M = \mathbb{H}^3$ is the 3-dimensional real hyperbolic space one has

$$K(t,x,y) = \frac{1}{(4\pi t)^{3/2}} \frac{d(x,y)}{\sinh d(x,y)} \exp\left(-t - \frac{d^2(x,y)}{4t}\right),$$

(cf. Section 3 in [23]).

Further information about heat kernels is contained in [14, 26] and in the survey article [37].

2.2. The Heat Semigroup on L^p-Spaces

In this section we denote by M an arbitrary complete Riemannian manifold with heat kernel K. Our aim is to give a definition of the Laplace-Beltrami operator $\Delta_{M,p}$ on the *complex* $L^p(M)$-spaces, $p \in [1,\infty)$, where we construct the $L^p(M)$-spaces with respect to the Riemannian measure of our manifold M.

The Laplace-Beltrami operator acting on the Hilbert space $L^2(M)$ with domain $C_c^\infty(M)$ (the set of differentiable functions with compact support) is, for complete manifolds M, essentially self-adjoint. That is to say, the closure of the Laplace-Beltrami operator, we call it also Δ_M, is self-adjoint. Furthermore, Δ_M is a positive operator, i.e. $\langle \Delta_M f, f\rangle \geq 0$ for $f \in \mathrm{dom}(\Delta_M)$, and therefore $-\Delta_M$ generates a bounded analytic semigroup $\mathrm{e}^{-t\Delta_M}$ on $L^2(M)$, which can be defined by the spectral theorem for unbounded self-adjoint operators. One can prove the equality

$$\mathrm{e}^{-t\Delta_M} u_0(x) = \int_M K(t,x,y) u_0(y)\,dvol(y),$$

for all initial data $u_0 \in L^2(M)$, where K denotes the corresponding heat kernel. Details for this can be found in [71].

Lemma 2.5. *The semigroup* $e^{-t\Delta_M}$ *on* $L^2(M)$ *is a* symmetric Markov semigroup, *i.e.*

(i) Δ_M *is positive and self-adjoint,*

(ii) $e^{-t\Delta_M}$ *is positive, which means* $e^{-t\Delta_M} f \geq 0$ *for all* $f \in L^2(M)$ *with* $f \geq 0$,

(iii) $e^{-t\Delta_M}$ *is a contraction on* $L^\infty(M)$.

Proof. It remains to prove (ii) and (iii). But these properties follow immediately from the positivity of the heat kernel K and the fact

$$\int_M K(t,x,y)\, dvol(y) \leq 1.$$

□

The terminology for these semigroups is not unique. Another widely used term is *submarkovian semigroup* reflecting better that these semigroups are L^∞-contractions. I would prefer to include the condition $e^{-t\Delta_M}1 = 1$ in the definition of symmetric Markov semigroups, which is in our case equivalent to stochastic completeness and therefore satisfied for the heat semigroups on locally symmetric spaces.

For a proof of the next theorem we refer to Theorem 1.4.1 and Theorem 1.4.2 in [19].

Theorem 2.6. (a) *The semigroup* $e^{-t\Delta_M}$ *leaves the set*

$$L^1(M) \cap L^\infty(M) \subset L^2(M)$$

invariant and $e^{-t\Delta_M}$ *may be extended to a positive contraction semigroup* $T_p(t)$ *on* $L^p(M)$ *for all* $p \in [1,\infty]$.

(b) *These semigroups are* consistent, *i.e.*

$$T_p(t)f = T_q(t)f$$

if $f \in L^p(M) \cap L^q(M)$, *and they are strongly continuous if* $p \in [1,\infty)$.

(c) *If* $p \in (1,\infty)$, *the semigroup* $T_p(t)$ *is a bounded analytic semigroup with angle of analyticity* $\theta_p \geq \frac{\pi}{2}\left(1 - |\frac{2}{p} - 1|\right)$.

The proof of part (c) in [19] relies on the Stein interpolation theorem. Another approach, which yields a better estimate of the angle θ_p, can be found in [59]: V.A. Liskevich and M.A. Perelmuter proved for the angle of analyticity θ_p of any submarkovian semigroup:

$$\theta_p \geq \frac{\pi}{2} - \arctan \frac{|p-2|}{2\sqrt{p-1}} \qquad p \in (1,\infty). \tag{2.3}$$

Notice that in general a symmetric Markov semigroup on $L^1(M)$ needs not be analytic. For an example where this happens cf. Section 4.3 in [19]. However, if M is a complete Riemannian manifold with bounded geometry, the heat semigroup $T_1(t) = \mathrm{e}^{-t\Delta_M} : L^1(M) \to L^1(M)$ is analytic in *some* sector (cf. [77, 18]).

Definition 2.7. *The Laplace-Beltrami operator $\Delta_{M,p}$ on $L^p(M)$, $p \in [1,\infty)$ is the negative of the generator of the strongly continuous contraction semigroup $T_p(t)$ on $L^p(M)$.*

In the subsequent sections we write Δ_M instead of $\Delta_{M,2}$ for the Laplace-Beltrami operator on $L^2(M)$ and $\mathrm{e}^{-t\Delta_{M,p}}$ instead of $T_p(t)$.

Lemma 2.8. *If $p, q \in [1,\infty)$, the operators $\Delta_{M,p}$ and $\Delta_{M,q}$ are consistent, i.e.*

$$\Delta_{M,p} f = \Delta_{M,q} f \qquad \textit{for any } f \in \mathrm{dom}(\Delta_{M,p}) \cap \mathrm{dom}(\Delta_{M,q}).$$

Proof. Since the semigroups $\mathrm{e}^{-t\Delta_{M,p}}$ and $\mathrm{e}^{-t\Delta_{M,q}}$ are consistent, we have for $f \in \mathrm{dom}(\Delta_{M,p}) \cap \mathrm{dom}(\Delta_{M,q})$:

$$\frac{1}{t}\left(\mathrm{e}^{-t\Delta_{M,p}} f - f\right) \xrightarrow{||\cdot||_{L^p}} -\Delta_{M,p} f \quad (t \downarrow 0)$$

and

$$\frac{1}{t}\left(\mathrm{e}^{-t\Delta_{M,p}} f - f\right) = \frac{1}{t}\left(\mathrm{e}^{-t\Delta_{M,q}} f - f\right) \xrightarrow{||\cdot||_{L^q}} -\Delta_{M,q} f \quad (t \downarrow 0).$$

Furthermore,

$$\Delta_{M,p} f - \Delta_{M,q} f \in L^p(M) + L^q(M)$$

and $L^p(M) + L^q(M)$ is a Banach space for the norm

$$||g||_{L^p+L^q} := \inf\left\{||h_1||_{L^p} + ||h_2||_{L^q} : h_1 \in L^p(M), h_2 \in L^q(M) \text{ with } g = h_1 + h_2\right\}.$$

In particular, we obtain

$$||\Delta_{M,p} f - \Delta_{M,q} f||_{L^p+L^q} \leq ||\frac{1}{t}\left(\mathrm{e}^{-t\Delta_{M,p}} f - f\right) + \Delta_{M,p} f||_{L^p} + ||\frac{1}{t}\left(\mathrm{e}^{-t\Delta_{M,q}} f - f\right) + \Delta_{M,q} f||_{L^q} \longrightarrow 0 \quad (t \downarrow 0).$$

□

If we identify as usual the dual space of $L^p(M)$, $p < \infty$, with $L^{p'}(M)$, $\frac{1}{p} + \frac{1}{p'} = 1$, we obtain the following lemma.

Lemma 2.9. *The adjoint (or dual) operator of $\Delta_{M,p}$ equals $\Delta_{M,p'}$:*

$$(\Delta_{M,p})' = \Delta_{M,p'}.$$

Proof. This follows from the facts that both operators are closed and coincide on $C_c^\infty(M)$: Since the adjoint of a closed operator is always closed, we only need to show the last claim. For this we take $f_1, f_2 \in C_c^\infty(M)$. Using Stokes' theorem and the formula $\mathrm{div}(fZ) = g(\mathrm{grad} f, Z) + f \mathrm{div} Z$ we obtain Greens formula

$$\int_M f_1 \Delta_M f_2 \, dvol(x) = \int_M g(\mathrm{grad} f_1, \mathrm{grad} f_2) \, dvol(x),$$

where g denotes the Riemannian metric on M. Denoting by $\langle \cdot, \cdot \rangle$ the duality bracket we obtain

$$\begin{aligned} \langle (\Delta_{M,p})' f_1, f_2 \rangle &= \langle f_1, \Delta_{M,p} f_2 \rangle \\ &= \int_M f_1 \Delta_{M,p} f_2 \, dvol(x) = \int_M (\Delta_{M,p'} f_1) f_2 \, dvol(x) \\ &= \langle \Delta_{M,p'} f_1, f_2 \rangle. \end{aligned}$$

Now the lemma follows from the Hahn-Banach theorem or by constructing certain cut-off functions f_2. □

Definition 2.10. *The L^p-spectrum $\sigma(\Delta_{M,p})$ of a Riemannian manifold is the spectrum of $\Delta_{M,p}$, $p \in [1, \infty)$.*

The theory of analytic semigroups and (2.3) imply that $\sigma(\Delta_{M,p})$, $p \in (1, \infty)$, is contained in the sector

$$\left\{ z \in \mathbb{C} \setminus \{0\} : |\arg(z)| \leq \frac{\pi}{2} - \theta_p \right\} \cup \{0\} \subset \left\{ z \in \mathbb{C} \setminus \{0\} : |\arg(z)| \leq \arctan \frac{|p-2|}{2\sqrt{p-1}} \right\} \cup \{0\},$$

which degenerates in the case $p = 2$ to the set of non-negative real numbers. Our considerations in Chapter 4 yield the result that the angle (2.3) is optimal with respect to the class of symmetric markov semigroups. Other proofs for this can be found in [53, 54, 78], where the argument in [54] even shows that the angle (2.3) is optimal for the class of Neumann Laplacians on domains in Euclidean space.

2.3. L^p-Spectrum of a Product Manifold

In this section we want to prove the following theorem.

Theorem 2.11. *Let $M = M_1 \times M_2$ be a (Riemannian) product of Riemannian manifolds M_1 and M_2. Furthermore, we denote by $\Delta_{M,p}, \Delta_{M_1,p}$, and $\Delta_{M_2,p}$ the Laplace-Beltrami operators on the respective L^p-spaces. Then the L^p-spectrum for $p \in (1, \infty)$ of $\Delta_{M,p}$ is the set-theoretic sum of the spectra of $\Delta_{M_1,p}$ and $\Delta_{M_2,p}$:*

$$\sigma(\Delta_{M,p}) = \sigma(\Delta_{M_1,p}) + \sigma(\Delta_{M_2,p}) = \{ w + z : w \in \sigma(\Delta_{M_1,p}), z \in \sigma(\Delta_{M_2,p}) \}.$$

Proof. We obtain for all $f \in C_c^\infty(M_1 \times M_2)$ as a consequence of representation (2.1) the formula

$$\Delta_{M,p} f(x_1, x_2) = \Delta_{M_1,p} f(\cdot, x_2)|_{x_1} + \Delta_{M_2,p} f(x_1, \cdot)|_{x_2}. \tag{2.4}$$

The statement of the theorem follows easily for $p = 2$ if the manifolds M_1 and M_2 are compact. Note that the spectrum of compact manifolds is discrete and consists only of eigenvalues, cf. Section 3.1 below. In this case we can find complete orthonormal systems $\{\Phi_i : i \in I\}$ of $L^2(M_1)$ and $\{\Psi_j : j \in J\}$ of $L^2(M_2)$ consisting of eigenfunctions for Δ_{M_1} and Δ_{M_2}. Then the set $\{\Phi_i \Psi_j : i \in I, j \in J\}$ is a complete orthonormal system of $L^2(M)$ consisting of eigenfunctions for Δ_M and the claim follows. For $p \in (1, \infty)$ the result follows since the L^p-spectrum of a compact Riemannian manifold does not depend on p. This is a consequence of a result due to K.-T. Sturm, cf. [72].
If the manifold is non-compact, the spectrum needs not be discrete and may depend on p. To prove the theorem in full generality we need deeper results from the theory of tensor products. For more details we refer to the Appendix A and the book [25]. The Banach space $L^p(M_1 \times M_2)$ is isometrically isomorphic to the tensor product $L^p(M_1) \tilde{\otimes}_{\Delta_p} L^p(M_2)$. Thus, we obtain in view of formula (2.4)

$$\Delta_{M,p} = \Delta_{M_1,p} \otimes I + I \otimes \Delta_{M_2,p}.$$

It is known, that there is a *uniform cross norm* (or *tensor norm*) g_p which coincides with the norm Δ_p on $L^p(M_1) \otimes L^p(M_2)$. Recall also that the semigroups $\mathrm{e}^{-t\Delta_{M_1,p}}$ and $\mathrm{e}^{-t\Delta_{M_2,p}}$ are analytic. Therefore, our theorem is a special case of Theorem 5 in [68] and the proof is complete. □

2.4. The Heat Kernel on a Symmetric Space of non-positive Curvature

The asymptotic behavior of the heat kernel for $t \to \infty$ on symmetric spaces X was investigated in a series of papers (cf. [1] and the references therein):

E.B. Davies and N. Mandouvalos determined the asymptotic behavior for all real hyperbolic spaces $X = \mathbb{H}^{n+1}$ with $n \geq 1$:

$$K(t, x, y) \asymp t^{-\frac{n+1}{2}} (1 + d + t)^{n/2-1} (1 + d) \exp\left(-\frac{1}{4} n^2 t - \frac{1}{2} n d - \frac{d^2}{4t}\right), \tag{2.5}$$

for all $t > 0$ and $x, y \in X$ such that $d = d(x, y)$.[1]

J.-P. Anker and L. Ji generalized this result to all symmetric spaces of non-positive curvature for those $t > 0$ and $x, y \in X$ such that the time variable t is large compared to the distance $d(x, y)$ between the points x and y (cf. [1]).

[1]We write $f \asymp h$ for functions f and h if there is a positive constant c such that $\frac{1}{c} h \leq f \leq c h$.

Finally, in [2] or rather [64] J.-P. Anker and P. Ostellari were able to give a proof without this additional assumption. Before we can state this result, we need some preparation.

We again identify the symmetric space X with G/K and use the terminology of Chapter 1. Notice first that G acts by isometries on X and that therefore the Laplace-Beltrami operator is G-invariant: $\Delta_X(f \circ g) = (\Delta_X f) \circ g$ for all $f \in C^\infty(X)$ and $g \in G$. The uniqueness of the fundamental solution of the heat equation (c.f. Theorem 2.2) then implies the G-invariance of the heat kernel, i.e. $K(t, gx, gy) = K(t, x, y)$ for all $g \in G$ and $x, y \in X$. If we denote again by x_0 the base point of X and if we choose points x, y in the homogeneous space X, there are isometries $g, h \in G$ such that $x = gx_0$ and $y = hx_0$. Because of the Cartan decomposition $G = K \exp \overline{\mathfrak{a}^+} K$ of the Lie group G (cf. Theorem 1.7) there are $k, k' \in K$ and $H = H(g^{-1}h) \in \overline{\mathfrak{a}^+}$ with $g^{-1}h = k \exp H(g^{-1}h)k'$. We can therefore write the heat kernel as follows:

$$\begin{aligned} K(t, x, y) &= K(t, gx_0, hx_0) = K(t, x_0, g^{-1}hx_0) \\ &= K(t, x_0, k \exp H(g^{-1}h)k'x_0) = K(t, x_0, \exp Hx_0). \end{aligned}$$

Of course, the isometries g and h are not necessarily uniquely determined. But $H \in \overline{\mathfrak{a}^+}$ is uniquely determined by x and $y \in X$: assuming the isometries g' and $h' \in G$ satisfy also $x = g'x_0$ and $y = h'x_0$, we clearly have $g' = gk_1$ and $h' = hk_2$ with $k_1, k_2 \in K$. On the other hand, this implies $g'^{-1}h' = k_1^{-1}g^{-1}hk_2 = k_1^{-1}k \exp H(g^{-1}h)k'k_2$ and the claim is proven because the $H \in \overline{\mathfrak{a}^+}$ in the Cartan decomposition is unique. For the distance between x and $y \in X$ we obtain by an analogous calculation the formula

$$d(x, y) = d(x_0, \exp Hx_0) = ||H||.$$

J.-P. Anker and P. Ostellari proved the following theorem.

Theorem 2.12. *For all $H \in \overline{\mathfrak{a}^+}$ and all $t > 0$ we have*

$$\begin{aligned} K(t, x_0, \exp Hx_0) &\asymp t^{-n/2} \left(\prod_{\alpha \in \Sigma_0^+} (1 + \langle \alpha, H \rangle)(1 + t + \langle \alpha, H \rangle)^{\frac{m_\alpha + m_{2\alpha}}{2} - 1} \right) \\ &\quad \cdot e^{-||\rho||^2 t - \langle \rho, H \rangle - \frac{||H||^2}{4t}}, \end{aligned}$$

where Σ_0^+ denotes the set of indivisible positive roots and ρ half the sum of the positive roots (counted according to their multiplicity).

We conclude this section by explaining how we can recover the heat kernel estimate (2.5) due to E.B. Davies and N. Mandouvalos for the hyperbolic plane $\mathbb{H}^2$ from the preceding theorem. We first remind the reader of the fact

$$\mathbb{H}^2 = SL(2,\mathbb{R})/SO(2,\mathbb{R}).$$

The Lie algebra $\mathfrak{sl}(2,\mathbb{R}) = \{Z \in \mathbb{R}^{2\times 2} : \operatorname{tr}(Z) = 0\}$ has dimension 3, and we choose the Cartan subalgebra

$$\mathfrak{a} := \left\{ \begin{pmatrix} \lambda & 0 \\ 0 & -\lambda \end{pmatrix} : \lambda \in \mathbb{R} \right\}.$$

Let us denote by E_{ij} the 2×2 matrix that has a 1 in the position (i,j) and zeros elsewhere. Because of $\operatorname{ad}(H)(Z) = HZ - ZH$, we obtain for $H := \begin{pmatrix} \lambda_1 & 0 \\ 0 & \lambda_2 \end{pmatrix} \in \mathfrak{a}$ the following:

$$\begin{aligned} \operatorname{ad}(H)(E_{ij}) &= (\lambda_i - \lambda_j)E_{ij} \qquad (i \neq j), \\ \operatorname{ad}(H)(E_{11} - E_{22}) &= 0. \end{aligned}$$

The only roots are therefore α and $-\alpha$ with $\alpha(\begin{pmatrix} \lambda_1 & 0 \\ 0 & \lambda_2 \end{pmatrix}) = \lambda_1 - \lambda_2$, and the corresponding root spaces are one dimensional, more precisely

$$\mathfrak{g}_\alpha = \operatorname{span}(E_{12}) \qquad \text{and} \qquad \mathfrak{g}_{-\alpha} = \operatorname{span}(E_{21}).$$

If we choose the Weyl chamber

$$\mathfrak{a}^+ = \left\{ \begin{pmatrix} \lambda & 0 \\ 0 & -\lambda \end{pmatrix} : \lambda > 0 \right\},$$

we have $\Sigma^+ = \Sigma_0^+ = \{\alpha\}$ and $\rho = \frac{1}{2}\alpha$. The inner product on $\mathfrak{sl}(2,\mathbb{R})$ induced from the Killing form is just twice the Euclidean inner product on $\mathbb{R}^{2\times 2}$:

$$\langle Z_1, Z_2 \rangle = 2\operatorname{tr}(Z_1 \cdot Z_2^\top).$$

Therefore $H_1 := \frac{1}{2}\begin{pmatrix} 1 & 0 \\ 0 & -1 \end{pmatrix}$ is the vector in $\mathfrak{a}^+$ with norm 1. We obtain

$$||\rho|| = \rho(H_1) = \frac{1}{2}\alpha(H_1) = \frac{1}{2}.$$

After these comments, the right hand side in Theorem 2.12 reads for general $H \in \overline{\mathfrak{a}^+}$ as follows:

$$t^{-1}(1 + ||H||)(1 + t + ||H||)^{-1/2} e^{-\frac{1}{4}t - \frac{1}{2}||H|| - \frac{||H||^2}{4t}}.$$

Because of $||H|| = d(x,y)$ this is just the right hand side of (2.5) in the case $n = 1$.

2.5. The Heat Kernel on Quotients of Cartan-Hadamard Manifolds

In this section we want to determine the heat kernel k on a quotient $M = \Gamma\backslash X$ of a Cartan-Hadamard manifold X by a discrete subgroup Γ of the isometry group that acts freely on X if the heat kernel K on the universal covering space X is given. Using Theorem 2.12 we will be able to estimate k in the next section if M is a locally symmetric space.
Recall, that a *Cartan-Hadamard manifold* is a connected, simply connected, complete Riemannian manifold with non-positive sectional curvature.

Theorem 2.13. *Let X denote an n-dimensional Cartan-Hadamard manifold. Assume, that the Ricci curvature is bounded from below by the constant $(n-1)\kappa < 0$. Furthermore, we take a discrete subgroup $\Gamma \subset \mathrm{Isom}(X)$ of the isometry group which acts freely on X. If we denote by K the heat kernel on X and by k the heat kernel on the quotient space $\Gamma\backslash X$, we have for all $x, y \in X$:*

$$k(t, \pi(x), \pi(y)) = \sum_{\gamma\in\Gamma} K(t, x, \gamma y)$$

with the covering map $\pi : X \to M$.

In the case of $X = \mathbb{H}^n$ there is a proof in [12]. If the quotient manifold M is compact and if the heat kernel K on X and the first derivatives of K satisfy certain estimates, this result can already be found in [27]. In [16] there is a proof for symmetric spaces of non-compact type.
In the remaining part of this section X and Γ are always assumed to satisfy the conditions of Theorem 2.13. But before we can give a proof of this theorem, we need to prove two lemmas.

Let $f \in C(X)$ be a Γ-invariant function, i.e. $f(\gamma x) = f(x)$ for all $\gamma \in \Gamma$ and $x \in X$. We define the projection Pf of f to $C(M)$ by $(Pf)(\pi(x)) := f(x)$. Because of the Γ-invariance of f the projection Pf is well-defined.

Lemma 2.14. *If $f \in C(X)$ is Γ-invariant and if the restriction of f to a fundamental domain $F \subset X$ for Γ has compact support ($f|_F \in C_c(F)$), we have:*

$$\mathrm{e}^{-t\Delta_M}(Pf) = P(\mathrm{e}^{-t\Delta_X} f).$$

Proof. This lemma will follow from the uniqueness theorem 2.2. We first consider the following two Cauchy problems on M or X respectively:

$$\begin{aligned} \frac{\partial}{\partial t}\tilde{u}_t(\tilde{x}) &= -\Delta_M \tilde{u}_t(\tilde{x}), \\ \tilde{u}_0(\tilde{x}) &= (Pf)(\tilde{x}), \end{aligned}$$

and

$$\begin{aligned} \frac{\partial}{\partial t} u_t(x) &= -\Delta_X u_t(x), \\ u_0(x) &= f(x). \end{aligned} \tag{2.6}$$

The solutions are

$$\tilde{u}_t(\tilde{x}) = (\mathrm{e}^{-t\Delta_M}(Pf))(\tilde{x})$$

and

$$u_t(x) = (\mathrm{e}^{-t\Delta_X} f)(x) = \int_X K(t,x,y) f(y)\, dvol(y).$$

Because of the observation

$$u_t(\gamma x) = \int_X K(t,x,\gamma^{-1}y) f(y)\, dvol(y) = \int_X K(t,x,y) f(\gamma y)\, dvol(y)$$

the solution $u_t(x)$ is also Γ-invariant. Now it suffices to show the equality $u_t = \tilde{u}_t \circ \pi$ since $u_t(x) = P(\mathrm{e}^{-\Delta_X} f)(\pi(x))$. For this we check whether $\tilde{u}_t \circ \pi$ is also a solution of the Cauchy problem (2.6):

$$-\Delta_X(\tilde{u}_t \circ \pi) = -(\Delta_M \tilde{u}_t) \circ \pi = \left(\frac{\partial}{\partial t}\tilde{u}_t\right) \circ \pi = \frac{\partial}{\partial t}(\tilde{u}_t \circ \pi),$$

where we used in the first step that the covering map $\pi : X \to M$ is a local isometry. Furthermore, we have:

$$\tilde{u}_0 \circ \pi(x) = (Pf)(\pi(x)) = f(x).$$

Because of $\int_X K(t,x,y)\, dvol(y) = 1$ (cf. Lemma 2.3) the boundedness of the solution $u_t(x)$ follows from the boundedness of f. An analogous argument implies also that $\tilde{u}_t \circ \pi(x)$ is bounded. But since bounded solutions to (2.6) are unique (cf. Theorem 2.2), the claim follows. □

Lemma 2.15. *Let* $N(s;x,y) := \#\{\gamma \in \Gamma : d(x,\gamma y) < s\}$ *denote the* orbit counting function. *Then there exists a constant* $C_1 > 0$*, only depending on* Γ*, the dimension* n *of* X *and* κ*, such that*

$$N(s;x,y) \leq C_1 \mathrm{e}^{(n-1)\sqrt{-\kappa}s}.$$

Proof. Choose $x, y \in X$ and a ball $B(y,\varepsilon)$ with center y and radius $\varepsilon = \varepsilon(\Gamma) > 0$ such that $B(y,\varepsilon) \cap B(\gamma y, \varepsilon) = \emptyset$ for all $\gamma \neq \mathrm{id}$. It follows

$$N(s;x,y) \mathrm{vol}\, B(y,\varepsilon) \leq \mathrm{vol}\, B(x, s+\varepsilon).$$

If we compare the volume of a ball in X with the volume of a comparison ball in the simply connected space form $\mathbb{M}^n_\kappa$ with constant sectional curvature $\kappa < 0$ or with the volume in the Euclidean space $\mathbb{R}^n$ respectively, we obtain the estimate (cf. Chapter 3.4 in [15]):

$$N(s;x,y) \leq \frac{\operatorname{vol} B(x,s+\varepsilon)}{\operatorname{vol} B(y,\varepsilon)} \leq \frac{c_{n,\kappa} e^{(n-1)\sqrt{-\kappa}(s+\varepsilon)}}{c_n' \varepsilon^n} = C_1 e^{(n-1)\sqrt{-\kappa} s}.$$

□

For symmetric spaces of non-compact type a sharper estimate is possible, c.f. for example [56]. The estimate therein is based on a sharper estimate of the volume of balls (cf. [50]).

Proof of Theorem 2.13. We first remark that

$$\tilde{k}(t,\pi(x),\pi(y)) := \sum_{\gamma\in\Gamma} K(t,x,\gamma y)$$

is a well-defined function $\tilde{k}$ on $(0,\infty)\times M\times M$ with values in $[0,\infty]$.

In the next step we show that the series converges uniformly on $[a,b]\times B\times B$ for all $0<a<b<\infty$ and any non-empty compact subset $B\subset X$. Therefore, $\sum_{\gamma\in\Gamma} K(t,x,\gamma y)$ is continuous on $(0,\infty)\times X\times X$.
We first claim that for any $R>0$ the set

$$\Gamma(B,R) := \{\gamma\in\Gamma : d(B,\gamma B)\leq R\}$$

is finite. To see this, we observe that $B_R := \{x\in X : d(B,x)\leq R+\operatorname{diam}(B)\}$ is a compact subset of X. Furthermore, we have $B\subset B_R$ and for each $\gamma\in\Gamma(B,R)$ also $\gamma B\subset B_R$. We conclude $\gamma B_R\cap B_R\neq\emptyset$. Since the subgroup Γ is discrete, the set $\Gamma(B,R)$ has to be finite.
Our aim is to prove that $\sum_{\gamma\in\Gamma\setminus\Gamma(B,R)} K(t,x,\gamma y)$ converges uniformly on $[a,b]\times B\times B$ to zero if $R\to\infty$. For this, it suffices to use the following upper bound of the heat kernel on X:

$$K(t,x,y)\leq C_2\, t^{-n/2}\exp\left(-\frac{d^2(x,y)}{2Dt}\right), \tag{2.7}$$

for all $t>0, D>2$ and a constant $C_2=C_2(D)>0$ that depends only on D. Such an estimate holds in all Cartan-Hadamard manifolds and a proof can be found in Section 7.4 of [37]. Combining this with the fact that the orbit counting function $N(s;x,y) := \#\{\gamma\in\Gamma : d(x,\gamma y)<s\}$ grows at most exponentially in s (cf. Lemma

2.15) we obtain:

$$\begin{aligned}\sum_{\gamma\in\Gamma\setminus\Gamma(B,R)} K(t,x,\gamma y) &\le C_2\, a^{-n/2} \sum_{\gamma\in\Gamma\setminus\Gamma(B,R)} \exp\left(-\frac{d^2(x,\gamma y)}{2Db}\right)\\ &\le C_2\, a^{-n/2} \sum_{m=1}^{\infty} \#\{\gamma\in\Gamma\setminus\Gamma(B,R) : mR \le d(x,\gamma y) < (m+1)R\}\\ &\qquad\cdot \exp\left(-\frac{m^2R^2}{2Db}\right)\\ &\le C_1C_2\, a^{-n/2} \sum_{m=1}^{\infty} \mathrm{e}^{(n-1)\sqrt{-\kappa}(m+1)R} \exp\left(-\frac{m^2R^2}{2Db}\right).\end{aligned}$$

The last series is uniformly convergent (with respect to $R > 1$), independent of $(t,x,y) \in [a,b] \times B \times B$ and converges to zero if $R \to \infty$.

Now, the theorem is a consequence of Lemma 2.14: On the one hand, we have for all positive, Γ-invariant, and continuous functions f with $f|_F \in C_c(F)$

$$\begin{aligned}P\left(\mathrm{e}^{-t\Delta_X} f\right)(\pi(x)) &= \mathrm{e}^{-t\Delta_X} f(x) = \int_X K(t,x,y)f(y)\; dvol(y)\\ &= \sum_{\gamma\in\Gamma}\int_{\gamma F} K(t,x,y)f(y)\; dvol(y)\\ &= \sum_{\gamma\in\Gamma}\int_{F} K(t,x,\gamma y)f(y)\; dvol(y)\\ &= \int_F \left(\sum_{\gamma\in\Gamma} K(t,x,\gamma y)\right) f(y)\; dvol(y).\end{aligned}$$

Note, that we used the monotone convergence theorem due to B. Levi for interchanging summation and integration. On the other hand we have

$$\left(\mathrm{e}^{-t\Delta_M}(Pf)\right)(\pi(x)) = \int_F k(t,\pi(x),\pi(y))f(y)\; dvol(y),$$

where k denotes the heat kernel on M. This concludes the proof since the mapping $\sum_{\gamma\in\Gamma} K(t,x,\gamma y)$ is continuous. □

Remark 2.16. The proof above shows that Theorem 2.13 holds for more general Riemannian manifolds if we have

(i) a rough heat kernel estimate like (2.7) and

(ii) an upper bound for the orbit counting function that grows at most exponentially, c.f. Lemma 2.15.

Chapter 3.

Heat Kernel Estimates and L^2-Spectrum

In this chapter we first give an upper bound for the heat kernel on general compact Riemannian manifolds. After that, we use the results from Sections 2.4 and 2.5 to prove (upper) Gaussian bounds of the heat kernel k on non-compact locally symmetric spaces $M := \Gamma\backslash X$. This generalizes results obtained by E.B. Davies and N. Mandouvalos in [23] for non-compact hyperbolic manifolds.

3.1. General Compact Riemannian Manifolds

In this section we denote by M always a compact, connected Riemannian manifold (without boundary). For details to the results below consult for example [14] or [70].

The resolvent $(\Delta_M + 1)^{-1} : L^2(M) \to L^2(M)$ of the Laplace-Beltrami operator $\Delta_M := -\operatorname{div}\operatorname{grad}$ has the integral kernel (Green function) $G_{-1}(x, y) = \int_0^\infty e^{-t} K(t, x, y)\, dt$ where $K : (0, \infty) \times M \times M \to \mathbb{R}$ denotes the fundamental solution of the heat equation on M. Since we assume M to be compact, we have $G_{-1} \in L^2(M \times M)$ and the resolvent $(\Delta_M + 1)^{-1}$ is therefore a Hilbert-Schmidt operator, in particular compact. Because of the spectral mapping theorem (for resolvents) we obtain: The L^2-spectrum of the Laplace-Beltrami operator $\Delta_M := -\operatorname{div}\operatorname{grad}$ on M is discrete and consists only of an infinite series of eigenvalues

$$0 = \lambda_0 < \lambda_1 \leq \lambda_2 \leq \cdots \to \infty,$$

where each eigenvalue appears according to its (finite) multiplicity.

Since the manifold M is compact, all the constant functions ($\neq 0$) are L^2-eigenfunctions and therefore λ_0 has to be zero. These are actually all the possible eigenfunctions for the eigenvalue 0 since any harmonic function on a compact and connected Riemannian manifold is a constant function.

Furthermore, we can find a complete orthonormal system of the Hilbert space $L^2(M)$ consisting of eigenfunctions for Δ_M. Elliptic regularity implies that every eigenfunction Φ is differentiable: $\Phi \in C^\infty(M)$.

If $K : (0, \infty) \times M \times M \to \mathbb{R}$ denotes again the heat kernel on M and $\{\Phi_j : j \in \mathbb{N}\}$ a complete orthonormal system of $L^2(M)$ consisting of eigenfunctions for the eigenvalues λ_j, $\Delta_M \Phi_j = \lambda_j \Phi_j$, we have

$$K(t,x,y) = \sum_{j=0}^{\infty} \mathrm{e}^{-\lambda_j t} \Phi_j(x)\Phi_j(y) = \frac{1}{\mathrm{vol}\,(M)} + \sum_{j=1}^{\infty} \mathrm{e}^{-\lambda_j t} \Phi_j(x)\Phi_j(y), \tag{3.1}$$

with uniform convergence on subsets of the form $[a, \infty) \times M \times M$, $a > 0$.

The next theorem shows that the heat kernel K on a compact Riemannian manifold converges for $t \to \infty$ exponentially in t and uniformly in $x, y \in M$ to the limit $\frac{1}{\mathrm{vol}\,(M)}$. Because of the physical interpretation of the heat equation and the heat kernel, this is plausible: The heat tries to diminish differences of temperature in M by flowing from regions of high temperature to regions of low temperature. Since M is compact, this will end up in a stationary state.

Theorem 3.1. *There is a constant $C = C(M) > 0$, only depending on M, such that for all $x, y \in M$ and $t \geq 1$ the following holds:*

$$\left| K(t,x,y) - \frac{1}{\mathrm{vol}\,(M)} \right| \leq C\mathrm{e}^{-\lambda_1 t}.$$

Proof. The manifold is compact. Therefore, the claim is an easy consequence of the observation

$$\begin{aligned}
\left| K(t,x,y) - \frac{1}{\mathrm{vol}\,(M)} \right| &= \left| \sum_{j=1}^{\infty} \mathrm{e}^{-\lambda_j t} \Phi_j(x)\Phi_j(y) \right| \\
&= \mathrm{e}^{-\lambda_1 t} \cdot \left| \sum_{j=1}^{\infty} \mathrm{e}^{-(\lambda_j - \lambda_1)t} \Phi_j(x)\Phi_j(y) \right| \\
&\leq \mathrm{e}^{-\lambda_1 t} \cdot \left(\sum_{j=1}^{N} \mathrm{e}^{-(\lambda_j - \lambda_1)t} |\Phi_j(x)\Phi_j(y)| \right. \\
&\qquad \left. + \left| \sum_{j=N+1}^{\infty} \mathrm{e}^{-(\lambda_j - \lambda_1)t} \Phi_j(x)\Phi_j(y) \right| \right).
\end{aligned}$$

Because of uniform convergence of the series (3.1) on $[1, \infty) \times M \times M$, we can find an $N \in \mathbb{N}$ such that the second term in parentheses is less than 1. Since M is compact, the first term is bounded and the claim follows. □

Gaussian (upper) bounds on compact Riemannian manifolds are not of major interest since the Gaussian factor $\exp(-\frac{d^2(x,y)}{4t})$ is bounded away from zero for fixed t. Hence, we will turn to (Gaussian) heat kernel estimates on non-compact locally symmetric spaces.

3.2. Poincaré Series and the Critical Exponent

Let us denote by $X = G/K$ as before a symmetric space of non-positive sectional curvature and by $\Gamma \subset G$ a discrete subgroup of the isometry group which acts freely on X. The resulting locally symmetric space is again denoted by $M = \Gamma\backslash X$.

A major role in our estimates and upper bounds plays the *Poincaré series*

$$P(s; x, y) := \sum_{\gamma\in\Gamma} \mathrm{e}^{-sd(x,\gamma y)}$$

with $s \in (0, \infty), x, y \in X$ and its *critical exponent*

$$\delta(\Gamma) := \inf\{s \in (0, \infty) \ : \ P(s; x, y) < \infty\}.$$

Since all $\gamma \in \Gamma$ are isometries, an application of the triangle inequality implies that the definition of the critical exponent $\delta(\Gamma)$ does not depend on the choice of the points x and $y \in X$. We further remark, that because of $P(s; \gamma_1 x, \gamma_2 y) = P(s; x, y)$ for all $\gamma_1, \gamma_2 \in \Gamma$, the Poincaré series $P(s; \cdot, \cdot)$ can be considered as a function on $M \times M$.

Recall the orbit counting function $N(R; x, y) := \#\{\gamma \in \Gamma : d(x, \gamma y) \leq R\}$. One can prove the equality

$$\delta(\Gamma) = \limsup_{R\to\infty} \frac{\log N(R; x, y)}{R}, \tag{3.2}$$

cf. [62] or [73]. The critical exponent $\delta(\Gamma)$ is therefore a measure for the exponential growth rate of Γ orbits in X.

Before we begin with estimating the Poincaré series, we give an upper bound for the critical exponent $\delta(\Gamma)$.

Lemma 3.2. *If ρ denotes (as above) half the sum of the positive roots, we have*

$$\delta(\Gamma) \leq 2||\rho||.$$

Proof. We consider the symmetric space $X = X_- \times \mathbb{R}^m$ with non-positive sectional curvature where X_- denotes a symmetric space of non-compact type. Then the following holds for the volume of a ball in X with center $x \in X$ and radius $R > 0$:

$$\operatorname{vol} B(x, R) \asymp R^m R^{\frac{(\operatorname{rank} X_-)-1}{2}} \mathrm{e}^{2||\rho||R}, \tag{3.3}$$

cf. [50]. As in the proof of Lemma 2.15 we obtain the estimate

$$N(R;x,y) \leq \frac{\operatorname{vol} B(x,R+\varepsilon)}{\operatorname{vol} B(y,\varepsilon)} \leq C\left(\frac{R+\varepsilon}{\varepsilon}\right)^m \left(\frac{R+\varepsilon}{\varepsilon}\right)^{\frac{(\operatorname{rank} X_-)-1}{2}} \mathrm{e}^{2||\rho||R}.$$

The claim now follows from formula (3.2). □

3.2.1. Estimates of the Poincaré Series

Since the Poincaré series appears in our heat kernel estimates, we prove in this subsection certain upper bounds for this series. These bounds are important if one is not only interested in the asymptotic behavior of the heat kernels $k(t,\tilde{x},\tilde{y})$ if $t \to \infty$ (for fixed points $\tilde{x}$ and $\tilde{y} \in M$) but also if the space variables $\tilde{x}$ and $\tilde{y}$ vary.

In the following lemma we denote by $\operatorname{inj}(\tilde{x})$ the *injectivity radius* of $\tilde{x} \in M = \Gamma\backslash X$. Recall the formula

$$\operatorname{inj}(\tilde{x}) = \frac{1}{2}\min\{d(x,\gamma x) : \gamma \in \Gamma \setminus \{\mathrm{id}\}\},$$

which is true for all $x \in X$ projecting to $\tilde{x}$, i.e. $\pi(x) = \tilde{x}$ where $\pi : X \to \Gamma\backslash X$ denotes the canonical projection. For such points x, we therefore define $\operatorname{inj}(x) := \operatorname{inj}(\tilde{x})$.

It turns out, that under the assumption $s > 2||\rho||$ it is easier to obtain upper bounds of the Poincaré series $P(s;x,y)$.

Lemma 3.3. *Let $s > 2||\rho||$ and choose $0 < 2\varepsilon < s-2||\rho||$. Then there is a constant $C = C(s,\varepsilon) > 0$, such that*

$$1 \leq P(s;x,x) \leq 1 + C\left(\frac{1}{\operatorname{inj}(x)}\right)^{m+\frac{(\operatorname{rank} X_-)-1}{2}} \cdot \mathrm{e}^{(2||\rho||-s+2\varepsilon)\operatorname{inj}(x)}.$$

Proof. The lower bound is trivial since $\mathrm{id} \in \Gamma$. The upper bound follows essentially from (3.3). In fact, we have

$$P(s;x,x) \leq \\ 1 + \sum_{n=0}^{\infty} \#\{\gamma \in \Gamma : \operatorname{inj}(x) + n \leq d(x,\gamma x) \leq \operatorname{inj}(x) + n + 1\} \cdot e^{(-s(\operatorname{inj}(x)+n))}.$$

Since the open balls $B(\gamma x, \operatorname{inj}(x))$ are pairwise disjoint, we obtain the following estimate for $S_n(x) := \#\{\gamma \in \Gamma : \operatorname{inj}(x) + n \leq d(x,\gamma x) \leq \operatorname{inj}(x) + n + 1\}$:

$$S_n(x)\operatorname{vol} B(x,\operatorname{inj}(x)) \leq \operatorname{vol} B(x, 2\operatorname{inj}(x) + n + 1).$$

For all $\varepsilon > 0$ we have $\operatorname{vol} B(x,R) \le c_\varepsilon e^{(2||\rho||+\varepsilon)R}$ with a constant $c_\varepsilon > 0$ that depends on the choice of ε. Using (3.3) we can conclude

$$\begin{aligned} S_n(x) &\le \frac{\operatorname{vol} B(x, 2\mathrm{inj}(x)+n+1)}{\operatorname{vol} B(x, \mathrm{inj}(x))} \\ &\le C_\varepsilon \left(\frac{1}{\mathrm{inj}(x)}\right)^{m+\frac{(\mathrm{rank}X_-)-1}{2}} \cdot e^{(2||\rho||+\varepsilon)(2\mathrm{inj}(x)+n+1)} e^{-2||\rho||\mathrm{inj}(x)}. \end{aligned}$$

This implies the following upper bound of the Poincaré series:

$$\begin{aligned} P(s;x,x) \le\ & 1 + C_\varepsilon \left(\frac{1}{\mathrm{inj}(x)}\right)^{m+\frac{(\mathrm{rank}X_-)-1}{2}} \cdot e^{(2||\rho||-s+2\varepsilon)\mathrm{inj}(x)} e^{2||\rho||+\varepsilon} \cdot \\ & \cdot \sum_{n=0}^{\infty} e^{(2||\rho||-s+\varepsilon)n}. \end{aligned}$$

Because of our choice of s and ε we have in particular $2||\rho|| - s + \varepsilon < 0$ and the geometric series $\sum_{n=0}^{\infty} e^{(2||\rho||-s+\varepsilon)n}$ equals $(1 - e^{2||\rho||-s+\varepsilon})^{-1}$. Now the proof is complete. □

Of course, it would suffice for the proof of the lemma above that ε satisfies the inequality $2||\rho|| - s + \varepsilon < 0$. But this (weaker) assumption does not guarantee that the term $e^{(2||\rho||-s+2\varepsilon)\mathrm{inj}(x)}$ converges (exponentially) to zero as $\mathrm{inj}(x) \to \infty$.

Recall, that a Riemannian Manifold M is said to have *bounded geometry* if its injectivity radius $\mathrm{inj}(M) := \inf_{x\in M} \mathrm{inj}(x)$ is bounded from below by a strictly positive constant and if its Ricci curvature is bounded from below. The second condition is always fulfilled if M is a locally symmetric space.

Corollary 3.4. *Let $M = \Gamma\backslash X$ be a locally symmetric space and choose $s > 2||\rho||$.*

(a) *Assume, that M has bounded geometry. Then the Poincaré series $P(s;x,x)$ is (for fixed s) bounded from above.*

(b) *If $x_n \in X$ is a sequence with $\mathrm{inj}(x_n) \to \infty$, it follows $P(s;x_n,x_n) \to 1$.*

For the following estimates of the Poincaré series we choose an arbitrary (but fixed) point $x' \in X$.

Lemma 3.5. *Choose $s > 2||\rho||$ and $x' \in X$. Then the following holds:*

(a) *There is a positive constant $C = C(x',s)$ (only depending on x' and s) such that*

$$P(s;x,x') \le C$$

for all $x \in X$.

(b) *There is a positive constant $C = C(x', s)$ (only depending on x' and s) such that*

$$P(s; x, x) \leq C e^{sd_M(\pi(x), \pi(x'))}$$

for all $x \in X$.

Proof. (a) The proof is similar to the preceding one:

$$\begin{aligned}
P(s; x, x') &\leq \sum_{n=1}^{\infty} \#\{\gamma \in \Gamma : n-1 \leq d(x, \gamma x') \leq n\} \cdot e^{-s(n-1)} \\
&\leq \sum_{n=1}^{\infty} \frac{\operatorname{vol} B(x, n + \operatorname{inj}(x'))}{\operatorname{vol} B(x', \operatorname{inj}(x'))} \cdot e^{-s(n-1)} \\
&\leq C_\varepsilon \left(\frac{1}{\operatorname{inj}(x')} \right)^{m + \frac{(\operatorname{rank} X_-)-1}{2}} \cdot e^{-2||\rho|| \operatorname{inj}(x')} \cdot \\
&\quad \cdot \sum_{n=1}^{\infty} e^{(2||\rho||+\varepsilon)(n+\operatorname{inj}(x'))} \cdot e^{-s(n-1)} \\
&= C_{\varepsilon, x', s} \sum_{n=1}^{\infty} e^{(2||\rho||+\varepsilon-s)n}.
\end{aligned}$$

If we choose ε sufficiently small, the last series converges and the claim follows.

(b) Using the triangle inequality $d(x, \gamma x) + d(x, \gamma' x') \geq d(\gamma x, \gamma' x')$ we can conclude for all $\gamma' \in \Gamma$:

$$\begin{aligned}
P(s; x, x) &\leq \sum_{\gamma \in \Gamma} e^{-sd(\gamma x, \gamma' x')} \cdot e^{sd(x, \gamma' x')} \\
&= P(s; x, \gamma' x') e^{sd(x, \gamma' x')} \\
&= P(s; x, x') e^{sd(x, \gamma' x')}.
\end{aligned}$$

We choose an isometry $\gamma' \in \Gamma$ with the property

$$d(x, \gamma' x') = \min_{\gamma \in \Gamma} d(x, \gamma x') = d_M(\pi(x), \pi(x')).$$

Now part (b) follows immediately from part (a).

□

3.3. Gaussian Bounds – Part 1

First of all we define

$$\rho_{min} := \min \left\{ \langle \rho, H \rangle \ : \ H \in \overline{\mathfrak{a}^+}, ||H|| = 1 \right\} \geq 0.$$

In the following we study the cases $\delta(\Gamma) < \rho_{min}$ and $\delta(\Gamma) \geq \rho_{min}$ separately. In the first case we obtain a Gaussian bound directly while we need in the second case a result due to A. Grigor'yan to obtain Gaussian bounds at least in certain situations.

Theorem 3.6. *Assume $\delta(\Gamma) < \rho_{min}$. Then there is for any $s \in (\delta(\Gamma), \rho_{min})$ a constant $C = C(s) > 0$ such that for all $t > 0$ and $\tilde{x}, \tilde{y} \in M = \Gamma\backslash X$ the estimate*

$$k(t, \tilde{x}, \tilde{y}) \leq Ct^{-n/2}(1+t)^m \exp\left(-||\rho||^2 t - \frac{d_M^2(\tilde{x}, \tilde{y})}{4t}\right) P(s; \tilde{x}, \tilde{y})$$

holds. Here, $m \geq 0$ is defined by $m := \sum_{\alpha \in \Sigma_0^+} \left(\frac{m_\alpha + m_{2\alpha}}{2} - 1\right) \geq 0$.

Proof. We use Theorem 2.12 to estimate $k(t, \tilde{x}, \tilde{y}) = \sum_{\gamma \in \Gamma} K(t, x, \gamma y)$, where $x, y \in X$ are chosen such that $\pi(x) = \tilde{x}$ and $\pi(y) = \tilde{y}$. For this we denote by $H(\gamma)$ the unique element in $\overline{\mathfrak{a}^+}$ with

$$K(t, x, \gamma y) = K(t, x_0, \exp H(\gamma) x_0)$$

and

$$d(x, \gamma y) = ||H(\gamma)||$$

(cf. Section 2.4). First, we obtain:

$$\begin{aligned} k(t, \tilde{x}, \tilde{y}) \;\leq\; & C_1 t^{-n/2}(1+t)^m \cdot \\ & \cdot \sum_{\gamma \in \Gamma} \left(\prod_{\alpha \in \Sigma_0^+} (1 + \langle \alpha, H(\gamma) \rangle)^{\frac{m_\alpha + m_{2\alpha}}{2}} \right) \mathrm{e}^{-||\rho||^2 t - \langle \rho, H(\gamma) \rangle - \frac{||H(\gamma)||^2}{4t}}. \end{aligned}$$

Because of $d_M(\tilde{x}, \tilde{y}) = \min_{\gamma \in \Gamma} d(x, \gamma y)$ it further follows:

$$\begin{aligned} k(t, \tilde{x}, \tilde{y}) \;\leq\; & C_1 t^{-n/2}(1+t)^m \mathrm{e}^{-||\rho||^2 t - \frac{d_M^2(\tilde{x}, \tilde{y})}{4t}} \cdot \\ & \cdot \sum_{\gamma \in \Gamma} \left(\prod_{\alpha \in \Sigma_0^+} (1 + \langle \alpha, H(\gamma) \rangle)^{\frac{m_\alpha + m_{2\alpha}}{2}} \right) \mathrm{e}^{-\langle \rho, H(\gamma) \rangle} \\ \leq\; & C_1 t^{-n/2}(1+t)^m \mathrm{e}^{-||\rho||^2 t - \frac{d_M^2(\tilde{x}, \tilde{y})}{4t}} \cdot \\ & \cdot \sum_{\gamma \in \Gamma} \left(\prod_{\alpha \in \Sigma_0^+} (1 + ||\alpha|| \cdot ||H(\gamma)||)^{\frac{m_\alpha + m_{2\alpha}}{2}} \right) \mathrm{e}^{-\rho_{min} ||H(\gamma)||}. \end{aligned}$$

Now we have a closer look at the last sum. Since the term

$$\prod_{\alpha \in \Sigma_0^+} (1 + ||\alpha|| \cdot ||H(\gamma)||)^{\frac{m_\alpha + m_{2\alpha}}{2}}$$

is the square root of a polynomial in $||H(\gamma)||$, we can find for every $s \in (\delta(\Gamma), \rho_{min})$ a constant $C_2 = C_2(s) > 0$, such that

$$\prod_{\alpha\in\Sigma_0^+} (1 + ||\alpha|| \cdot ||H(\gamma)||)^{\frac{m_\alpha + m_{2\alpha}}{2}} \mathrm{e}^{-\rho_{min}||H(\gamma)||} \leq C_2 \mathrm{e}^{-s||H(\gamma)||} = C_2 \mathrm{e}^{-sd(x,\gamma y)}.$$

This concludes the proof. □

The condition $s < \rho_{min}$ prevents an application of the results from Subsection 3.2.1: Using the triangle inequality, we can conclude $P(s; x, y) \leq P(s; x, x)\mathrm{e}^{sd(x,y)}$. But for the proof of the estimates of the Poincaré series $P(s; x, x)$ in the mentioned subsection we made the assumption $s > 2||\rho||$. Therefore, we give in Section 3.5 further heat kernel estimates where this problem does not occur.

In the following we give an estimate of the heat kernel on quotients $M = \Gamma\backslash X$ for larger subgroups Γ, i.e. $\delta(\Gamma) \geq \rho_{min}$. The statement of the next theorem is also true for subgroups with $\delta(\Gamma) < \rho_{min}$ but the estimate is weaker than the one obtained in Theorem 3.6.

Theorem 3.7. *Assume $\delta(\Gamma) \geq \rho_{min}$. Then there is for all $\varepsilon > 0$ a constant $C = C(\varepsilon) > 0$ such that the following estimate for the heat kernel k on M holds:*

$$k(t, \tilde{x}, \tilde{y}) \leq Ct^{-n/2}(1 + t)^m \mathrm{e}^{-(||\rho||^2 - (\delta(\Gamma) - \rho_{min} + \varepsilon)^2)t} \cdot P(\delta(\Gamma) + \varepsilon/2; \tilde{x}, \tilde{y}),$$

where $m \geq 0$ is defined as in Theorem 3.6.

Proof. In order to estimate

$$k(t, \tilde{x}, \tilde{y}) = \sum_{\gamma\in\Gamma} K(t, x, \gamma y),$$

we use again Theorem 2.12. First of all, we concentrate on the term

$$\mathrm{e}^{-||\rho||^2 t - \langle\rho, H\rangle - \frac{||H||^2}{4t}} \leq \mathrm{e}^{-||\rho||^2 t - \rho_{min}||H|| - \frac{||H||^2}{4t}}.$$

A straightforward calculation shows that for any $\beta \in \mathbb{R}$ the right hand side of this inequality coincides with the left hand side of the next inequality:

$$\mathrm{e}^{-(\rho_{min}+\beta)||H|| - ||\rho||^2 t} \mathrm{e}^{-(\frac{||H||}{2\sqrt{t}} - \beta\sqrt{t})^2} \mathrm{e}^{\beta^2 t} \leq \mathrm{e}^{-(\rho_{min}+\beta)||H|| - ||\rho||^2 t} \mathrm{e}^{\beta^2 t}.$$

Choose $\varepsilon > 0$ and define $\beta := \beta(\varepsilon) := \delta(\Gamma) - \rho_{min} + \varepsilon$.
We obtain the estimate (cf. the proof of Theorem 3.6)

$$\begin{aligned} k(t, \tilde{x}, \tilde{y}) &\leq C_1 t^{-n/2}(1 + t)^m \mathrm{e}^{-(||\rho||^2 - (\delta(\Gamma) - \rho_{min} + \varepsilon)^2)t} \cdot \\ &\quad \cdot \sum_{\gamma\in\Gamma} \left(\prod_{\alpha\in\Sigma_0^+} (1 + \langle\alpha, H(\gamma)\rangle)^{\frac{m_\alpha + m_{2\alpha}}{2}} \right) \mathrm{e}^{-(\delta(\Gamma)+\varepsilon)||H(\gamma)||} \\ &\leq Ct^{-n/2}(1 + t)^m \mathrm{e}^{-(||\rho||^2 - (\delta(\Gamma) - \rho_{min} + \varepsilon)^2)t} \cdot P(\delta(\Gamma) + \varepsilon/2; x, y). \end{aligned}$$

In the last step we again used $\langle\alpha, H(\gamma)\rangle \leq ||\alpha|| \cdot ||H(\gamma)||$ and $||H(\gamma)|| = d(x, \gamma y)$. □

Remark 3.8. Because of $k \to 0$ (if $t \to \infty$) this estimate of the heat kernel k is only interesting if there is an $\varepsilon > 0$ such that $||\rho||^2 - (\delta(\Gamma) - \rho_{min} + \varepsilon)^2$ is positive. But this is equivalent to $||\rho|| + \rho_{min} > \delta(\Gamma) \geq \rho_{min}$.
For symmetric spaces $X = G/K$ of rank 1 the Lie subalgebra $\mathfrak{a}$ has dimension one and we therefore have $\rho_{min} = ||\rho||$. The condition from above reads in this case $||\rho|| \leq \delta(\Gamma) < 2||\rho||$. If the Lie group G has additionally *Kazhdan's property (T)* this condition is satisfied by a discrete subgroup Γ if and only if $||\rho|| \leq \delta(\Gamma)$ and the co-volume $\mathrm{vol}\,(\Gamma\backslash X)$ is infinite (cf. Theorem 4.4 in [16]).

If we apply the following result due to A. Grigory'an, we obtain for all subgroups Γ with $\rho_{min} \leq \delta(\Gamma) < ||\rho|| + \rho_{min}$ a Gaussian estimate.

Theorem 3.9 (cf. [36] or [37]). *Choose two points x and y on a connected Riemannian manifold M and denote by k the heat kernel on M. Assume, we have for all $t \in (0, T)$ (with T being a positive real number or infinity)*

$$k(t, x, x) \leq \frac{1}{f(t)}$$

and

$$k(t, y, y) \leq \frac{1}{g(t)}$$

with regular *functions f and g (compare the definition below). Then there is for all $D > 4$ a constant $\mu = \mu(D) > 0$ such that the following estimate holds for any $t \in (0, T)$ and all $x, y \in M$:*

$$k(t, x, y) \leq \frac{const.}{\sqrt{f(\mu t) g(\mu t)}} \exp\left(-\frac{d^2(x, y)}{Dt}\right).$$

Definition 3.10. *We call a function $f : (0, \infty) \to \mathbb{R}$* regular *if it satisfies the conditions* (i) *and* (ii):

(i) *The function f is positive and monotone increasing.*

(ii) *There are constants $A \geq 1$ and $a > 1$, such that the following inequality holds for all $0 < t_1 < t_2$:*

$$\frac{f(at_1)}{f(t_1)} \leq A \frac{f(at_2)}{f(t_2)}.$$

By using the estimates from Theorem 3.7 and Theorem 3.9 we obtain the corollary announced above:

Corollary 3.11. *Assume $\rho_{min} \leq \delta(\Gamma) < \rho_{min} + ||\rho||$. Then we have for the heat kernel k on $M = \Gamma\backslash X$ the Gaussian estimate*

$$\begin{aligned} k(t, \tilde{x}, \tilde{y}) \ \leq \ & Ct^{-n/2}(1+t)^m \mathrm{e}^{-(||\rho||^2 - (\delta(\Gamma) - \rho_{min} + \varepsilon)^2)\mu t} \cdot \\ & \mathrm{e}^{-\frac{d_M^2(\tilde{x}, \tilde{y})}{Dt}} \sqrt{P(\delta(\Gamma) + \varepsilon/2; \tilde{x}, \tilde{x})} \sqrt{P(\delta(\Gamma) + \varepsilon/2; \tilde{y}, \tilde{y})}, \end{aligned}$$

with constants $D > 4, \mu = \mu(D) > 0, C = C(\varepsilon, D) > 0$ and a constant $\varepsilon > 0$ small enough, such that the term $||\rho||^2 - (\delta(\Gamma) - \rho_{min} + \varepsilon)^2$ is positive.

Proof. We define for each $\tilde{x} \in M$ a function $f : (0, \infty) \to \mathbb{R}$ by (cf. Theorem 3.7):

$$\frac{1}{f(t)} := Ct^{-n/2}(1+t)^m \mathrm{e}^{-\left(||\rho||^2 - (\delta(\Gamma) - \rho_{min} + \varepsilon)^2\right)t} \cdot P(\delta(\Gamma) + \varepsilon/2; \tilde{x}, \tilde{x}).$$

Now, the corollary is a direct consequence of Theorem 3.9 if f is a regular function. If we choose ε as mentioned in the corollary we see immediately that condition (i) is fulfilled. To check property (ii), we first prove the existence of a $T > 0$ such that for $t \geq T$ the quotient $\frac{f(at)}{f(t)}$ is monotone increasing. For $t \leq T$ we can show that there is a constant $A > 1$ with $f(at) \leq Af(t)$. The regularity of f now follows as in [36], p.38. □

In this case we also have the problem that $\delta(\Gamma) + \varepsilon/2$ is smaller than $2||\rho||$ and therefore, we cannot apply the estimates of the Poincaré series obtained in the previous section. This problem will be remedied in Section 3.5 under the additional assumption $\dim(X) \geq 3$, too.

3.4. Heat Kernels and L^2-Spectrum

The results of the preceding section can be applied in order to give a lower bound for the *bottom of the L^2-spectrum*

$$\lambda_0(M) := \inf\{\lambda : \lambda \in \sigma(\Delta_M)\} \geq 0$$

for locally symmetric spaces $M := \Gamma\backslash X$. If M is a Riemannian manifold with finite volume, we always have $\lambda_0(M) = 0$ since the constant functions are contained in $L^2(M)$. The basis for our estimates is the following lemma.

Lemma 3.12. *Let M be a non-compact, connected Riemannian manifold with Laplace-Beltrami operator Δ_M and heat kernel K. Assume, there are $b > 0, E \in \mathbb{R}$ and $x \in M$ with*

$$K(t, x, x) \leq b\mathrm{e}^{-Et}$$

for all $t \geq 1$. Then it follows $\lambda_0(M) \geq E$.

A similar result can be found in [23] (Lemma 5.3) for hyperbolic manifolds with stronger assumptions: The authors require such an estimate of the heat kernel for *all* points $(x, y) \in M \times M$. In the proof they use the spectral theorem for unbounded self-adjoint operators. We give a more elementary proof by recalling the construction of heat kernels on Riemannian manifolds.

Definition 3.13. *With $\Omega \subset\subset M$ we denote connected open subsets of M whose closure $\overline{\Omega}$ is compact and whose boundary $\partial\Omega$ is a differentiable submanifold.*

We consider for $\Omega \subset\subset M$ the Dirichlet eigenvalue problem $\Delta_\Omega u = \lambda u$, $u|_{\partial\Omega} = 0$ in $L^2(\Omega)$. Since the subsets Ω are pre-compact, the spectrum of the Laplace-Beltrami operator Δ_Ω is discrete: $\sigma(\Delta_\Omega) = \{\lambda_n(\Omega) : n \in \mathbb{N}\}$ with $0 \leq \lambda_0(\Omega) \leq \lambda_1(\Omega) \leq ... \to \infty$. The associated heat kernel K_Ω is given by

$$K_\Omega(t, x, y) = \sum_{j=0}^{\infty} \mathrm{e}^{-\lambda_j(\Omega)t} \Phi_{j,\Omega}(x)\Phi_{j,\Omega}(y),$$

where the $\Phi_{j,\Omega}$ form an orthonormal basis consisting of eigenfunctions of Δ_Ω with eigenvalues $\lambda_j(\Omega)$. The maximum principle implies that for subsets $\Omega, \Omega' \subset\subset M$ with $\Omega \subset \Omega'$ the respective heat kernels satisfy the inequality $K_\Omega(t, x, y) \leq K_{\Omega'}(t, x, y)$.
For the remaining part of this section we denote by $(\Omega_k)_{k\in\mathbb{N}}$ a sequence of pre-compact subsets $\Omega_k \subset\subset M$ with $\overline{\Omega_k} \subset \Omega_{k+1}$ and $\cup_{k=1}^{\infty} \Omega_k = M$. One can prove that the heat kernel K on M is the limit of the heat kernels K_{Ω_k}:

$$K(t, x, y) = \lim_{k\to\infty} K_{\Omega_k}(t, x, y),$$

and the convergence is uniform on compact subsets of $(0, \infty) \times M \times M$. Details for this can be found in [14],[26] or [37].

Lemma 3.14. *For a Riemannian manifold M the following holds:*

$$\begin{aligned} \lambda_0(M) &= \inf\{(\Delta_M f, f)_{L^2(M)} : f \in C_c^\infty(M) \text{ with } ||f||_{L^2(M)} = 1\} \\ &= \inf\{\lambda_0(\Omega) : \Omega \subset\subset M\} \\ &= \lim_{k\to\infty} \lambda_0(\Omega_k). \end{aligned}$$

Proof. The first equality follows as in [45] since the spectrum of self-adjoint operators coincides always with the approximate point spectrum and since $C_c^\infty(M)$ is dense in the domain of the Laplace-Beltrami operator.
The second and third equality are consequences of the first one: On the one hand we have

$$\lambda_0(\Omega_k) = \inf\{(\Delta_{\Omega_k} f, f) : f \in C_c^\infty(\Omega_k), ||f||_{L^2(\Omega_k)} = 1\} \geq \lambda_0(M),$$

since $C_c^\infty(\Omega_k)$ is dense in the domain of the Dirichlet-Laplace operator Δ_{Ω_k}.
Choose on the other hand $\varepsilon > 0$ and $f \in C_c^\infty(M), ||f||_{L^2(M)} = 1$ with $(\Delta_M f, f) \leq \lambda_0(M)+\varepsilon$. If we choose $k \in \mathbb{N}$ with $\mathrm{supp} f \subset \Omega_k$, we obtain $\lambda_0(\Omega_k) \leq \lambda_0(M)+\varepsilon$. □

Now we are ready to give a proof of Lemma 3.12:

Proof of Lemma 3.12. We have $K(t,x,x) \leq be^{-Et}$ for all $t \geq 1$. First we deduce

$$\sum_{j=0}^{\infty} e^{-\lambda_j(\Omega_k)t}\Phi_{j,\Omega_k}^2(x) = K_{\Omega_k}(t,x,x) \leq be^{-Et}$$

for all $k \in \mathbb{N}$. This is equivalent to

$$\sum_{j=0}^{\infty} e^{(E-\lambda_j(\Omega_k))t}\Phi_{j,\Omega_k}^2(x) \leq b.$$

We choose $k_0 \in \mathbb{N}$ such that $x \in \Omega_k$ for all $k \geq k_0$. For these k we have the inequality $\lambda_0(\Omega_k) \geq E$ since $\Phi_{0,\Omega_k}(x) \neq 0$ (cf. [70], Lemma VI.3.10). An application of Lemma 3.14 concludes the proof. □

3.4.1. L^2-Spectrum of Locally Symmetric Spaces

Theorem 3.15. *Let $\Gamma\backslash X$ be a non-compact locally symmetric space. Then we have the following lower bounds for the bottom $\lambda_0(\Gamma\backslash X)$ of the L^2-spectrum:*

(a) $\lambda_0(\Gamma\backslash X) \geq ||\rho||^2$ *if* $\delta(\Gamma) < \rho_{min}$,

(b) $\lambda_0(\Gamma\backslash X) \geq ||\rho||^2 - (\delta(\Gamma) - \rho_{min})^2$ *if* $\rho_{min} \leq \delta(\Gamma) \leq ||\rho|| + \rho_{min}$.

Proof. The assertions follow from Lemma 3.12 if we recall Theorem 3.6 and Theorem 3.7: The function $h : [1,\infty) \to \mathbb{R}, t \to t^{-n/2}(1+t)^m$ is monotone decreasing since $m < n/2$. Therefore, we obtain in the first case the estimate

$$k(t,\tilde{x},\tilde{x}) \leq b(\tilde{x})e^{-||\rho||^2 t}$$

for all $t \geq 1$ with a positive function b on $\Gamma\backslash X$. In the second case, analogous considerations lead for any $\epsilon > 0$ to

$$k(t,\tilde{x},\tilde{x}) \leq b_\varepsilon(\tilde{x})e^{-(||\rho||^2-(\delta(\Gamma)-\rho_{min}+\varepsilon)^2)t}$$

for all $t \geq 1$. □

In case of $\delta(\Gamma) > ||\rho|| + \rho_{min}$ the term $||\rho||^2 - (\delta(\Gamma) - \rho_{min})^2$ is negative. Thus, we still have zero as lower bound for $\lambda_0(\Gamma\backslash X)$ in this case.

Remark 3.16. As in [23] ("note" after Theorem 5.1) we obtain also an upper bound for all locally symmetric spaces $\Gamma\backslash X$: $\lambda_0(\Gamma\backslash X) \leq ||\rho||^2$.

With this remark we can conclude:

Corollary 3.17. *If $\Gamma\backslash X$ is a locally symmetric space with $\delta(\Gamma) < \rho_{min}$, we have*

$$\lambda_0(\Gamma\backslash X) = ||\rho||^2.$$

The lower bounds for the bottom of the L^2-spectrum from above generalize numerous former achievements: If Γ is a Fuchsian group, the results can already be found in [31], [32], [33] and [65]. For hyperbolic spaces $X = \mathbb{H}^n$ with $n \geq 3$ these results are contained in [74]. K. Corlette proved these results for rank-1 symmetric spaces X of non-compact type (cf. [16]). A generalization to symmetric spaces of non-compact type with arbitrary rank is due to E. Leuzinger (cf. [57]). A closer look at the last paper shows that the proof therein also works in our case, i.e. for symmetric spaces $X = G/K$ with non-positive curvature and discrete subgroups $\Gamma \subset G$ of the reductive Lie group G which act freely on X. But the methods in the cited paper are different: The author uses Green function estimates due to J.-P. Anker and L. Ji, and these are also valid in our (more general) situation (cf. [1]).

We summarize all the results from above:

Theorem 3.18. *Let $X = G/K$ be a symmetric space of non-positive sectional curvature and $\Gamma \subset G$ a discrete subgroup of the reductive Lie group G that acts freely on X. Then the following holds:*

(a) $\lambda_0(\Gamma\backslash X) = ||\rho||^2$ *if* $0 \leq \delta(\Gamma) \leq \rho_{min}$,

(b) $||\rho||^2 - (\delta(\Gamma) - \rho_{min})^2 \leq \lambda_0(\Gamma\backslash X) \leq ||\rho||^2$ *if* $\rho_{min} \leq \delta(\Gamma) \leq ||\rho||$,

(c) $\max\{0, ||\rho||^2 - (\delta(\Gamma) - \rho_{min})^2\} \leq \lambda_0(\Gamma\backslash X) \leq ||\rho||^2 - (\delta(\Gamma) - ||\rho||)^2$
if $||\rho|| \leq \delta(\Gamma) \leq 2||\rho||$.

3.5. Gaussian Bounds – Part 2

In this section we want to apply a theorem due to E.B. Davies and N. Mandouvalos in order to obtain Gaussian bounds for a larger class of locally symmetric spaces. Furthermore, we also give new upper bounds of the heat kernel on $\Gamma\backslash X$ for subgroups $\Gamma \subset G$ with $\delta(\Gamma) < \rho_{min}$ which are suitable for applying the estimates of the Poincaré series of Section 3.2.1.

Theorem 3.19 (E.B. Davies & N. Mandouvalos, cf. [22]). *Let M denote a non-compact Riemannian manifold with dimension $n \geq 3$. We further denote by $\sigma : M \to (0, \infty)$ a C^∞ function and by $F \in \mathbb{R}$ a constant such that*

$$-\frac{\Delta_M \sigma}{\sigma} \geq F.$$

Assume that for all $t \in (0, 1]$ and $x \in M$ the following (on-diagonal) estimate for the heat kernel K on M holds:

$$K(t, x, x) \leq Ct^{-n/2}\sigma^2(x).$$

Then we have for all $\mu \in (0,1), t > 0$, and $x, y \in M$ the Gaussian estimate

$$K(t,x,y) \leq C_\mu t^{-n/2}\sigma(x)\sigma(y)\exp\left((2\mu - \lambda_0(M))t - \frac{d^2(x,y)}{4(1+\mu)t}\right).$$

We begin with the definition of a function σ on our symmetric space $X = G/K$ which descends to a suitable function on the quotient space $M = \Gamma\backslash X$ for a discrete subgroup $\Gamma \subset G$ that acts freely on X.

Definition 3.20. *Choose a non-negative function $f \in C_c^\infty([0,\infty))$ with $f(0) \neq 0$ and put $h : X \times X \to [0,\infty), (x,y) \mapsto f(d^2(x,y))$. For $s > \delta(\Gamma)$ we define*

$$\sigma : X \to (0,\infty),\ x \mapsto \int_X h(x,y)\sqrt{P(s;y,y)}\, dvol(y).$$

To show that this function has the properties we need (in view of Theorem 3.19) we prove the next lemma.

Lemma 3.21. *The function σ is differentiable, Γ-invariant, and defines therefore a function on the quotient space $\Gamma\backslash X$. Furthermore, we have:*

(a) *There is a constant $c > 1$ such that*

$$\frac{1}{c}\sqrt{P(s;x,x)} \leq \sigma(x) \leq c\sqrt{P(s;x,x)}.$$

(b) *There is a constant F with*

$$|\Delta_X\sigma(x)| \leq F\sigma(x).$$

In particular, we have $-\frac{\Delta_X\sigma}{\sigma} \geq -F$.

Proof. The differentiability of σ and the Γ-invariance are evidently clear. For the proof of the remaining assertions we first remark that the triangle inequality implies

$$P(s;y,y) \leq e^{2sd(x,y)}P(s;x,x).$$

(a) The definition of the function σ implies the existence of a constant $\beta > 0$, such that $h(x,y) = 0$ for all points $x, y \in X$ with $d(x,y) > \beta$. We therefore obtain

$$\begin{aligned}\sigma(x) &= \int_{d(x,y)\leq\beta} h(x,y)\sqrt{P(s;y,y)}\, dvol(y)\\ &\leq \int_{d(x,y)\leq\beta} h(x,y)e^{sd(x,y)}\sqrt{P(s;x,x)}\, dvol(y)\\ &\leq (\max h)e^{s\beta}\sqrt{P(s;x,x)}\mathrm{vol}\, B(x,\beta) \leq c_1(s,h)\sqrt{P(s;x,x)},\end{aligned}$$

with a constant $c_1 > 0$ that depends only on s and the function h. Notice, that we used in the last step the fact that the volume of a ball $B(x, \beta)$ in X is smaller than the volume of a comparison ball in some hyperbolic space $\mathbb{H}^n$ of constant curvature.
We choose $0 < a < \infty$ with $f(s) > 0$ for all $s \in [0, a^2]$. It follows:

$$\begin{aligned}
\sigma(x) &\geq \int_{d(x,y)\leq a} h(x,y) \mathrm{e}^{-sd(x,y)} \sqrt{P(s;x,x)}\, dvol(y) \\
&\geq \mathrm{e}^{-sa} \sqrt{P(s;x,x)} \int_{d(x,y)\leq a} h(x,y)\, dvol(y) \\
&\geq \mathrm{e}^{-sa} \sqrt{P(s;x,x)} \min_{s\in[0,a^2]} f(s) \int_{d(x,y)\leq a} dvol(y) \\
&\geq c_2 \sqrt{P(s;x,x)},
\end{aligned}$$

with a positive constant c_2. In the last step we applied again a volume comparison theorem in order to find a positive lower bound of the integral. More precisely, we compared the volume of the ball $B(x, a) \subset X$ with the volume of a Euclidean comparison ball.

(b) Using (a), we obtain

$$\begin{aligned}
|\Delta_X \sigma(x)| &= \left| \int_{d(x,y)\leq \beta} (\Delta_X h)(x,y) \sqrt{P(s;y,y)}\, dvol(y) \right| \\
&\leq \mathrm{e}^{s\beta} \sqrt{P(s;x,x)} \max(|\Delta_X h|) = c_3 \sqrt{P(s;x,x)} \\
&\leq F\sigma(x),
\end{aligned}$$

and in particular $F \geq \frac{|\Delta_X \sigma(x)|}{\sigma(x)} \geq \frac{\Delta_X \sigma(x)}{\sigma(x)}$.

□

This yields the

Corollary 3.22. *Let* $\dim X \geq 3$ *and* $\mu \in (0,1)$. *Then we obtain the following upper bounds for the heat kernel* k *on* $M := \Gamma\backslash X$:

(a) *If* $\delta(\Gamma) < \rho_{min}$ *and* $s > \delta(\Gamma)$:

$$\begin{aligned}
k(t, \tilde{x}, \tilde{y}) &\leq C_\mu t^{-n/2} \exp\left((2\mu - ||\rho||^2)t - \frac{d_M^2(\tilde{x}, \tilde{y})}{4(1+\mu)t} \right) \cdot \\
&\quad \cdot \sqrt{P(s; \tilde{x}, \tilde{x})} \sqrt{P(s; \tilde{y}, \tilde{y})}.
\end{aligned}$$

(b) *If* $\delta(\Gamma) \geq \rho_{min}$:

$$\begin{aligned}
k(t, \tilde{x}, \tilde{y}) &\leq C_{\varepsilon,\mu} t^{-n/2} \exp\left((2\mu - \lambda_0(M))t - \frac{d_M^2(\tilde{x}, \tilde{y})}{4(1+\mu)t} \right) \cdot \\
&\quad \cdot \sqrt{P(\delta(\Gamma) + \varepsilon; \tilde{x}, \tilde{x})} \sqrt{P(\delta(\Gamma) + \varepsilon; \tilde{y}, \tilde{y})},
\end{aligned}$$

for all $\varepsilon > 0$.

Proof. The assertions follow from Theorem 3.19 and Lemma 3.21 since the results in Section 3.3 imply in both cases a heat kernel estimate of the form

$$k(t,\tilde{x},\tilde{x}) \leq Ct^{-n/2}\sigma^2(x), \qquad t \in (0,1],$$

with some $x \in X$ such that $\pi(x) = \tilde{x}$.
We provide some details in case (b): Using Theorem 3.7 and Lemma 3.21 we conclude for $t \in (0,1]$:

$$k(t,\tilde{x},\tilde{x}) \leq C_\varepsilon t^{-n/2} P(\delta(\Gamma)+\varepsilon/2;\tilde{x},\tilde{x}) \leq C'_\varepsilon t^{-n/2}\sigma^2(x),$$

where we put $s := \delta(\Gamma)+\varepsilon/2$ in the definition of σ. The claim follows since $\sigma^2(x) \leq cP(\delta(\Gamma)+\varepsilon/2;x,x)$. □

Remark 3.23. These bounds contain also in case of $\delta(\Gamma) < \rho_{min}$ the functions $P(s;x,x)$ instead of $P(s;x,y)$ and s can be chosen as large as one wishes. Since there is also no restriction on the choice of $\varepsilon > 0$ in part (b), and since we estimated the functions $P(s;x,x)$ in Section 3.2.1 for $s > 2||\rho||$, we now have "complete" upper bounds for the heat kernels.

Using the estimate $\lambda_0(M) \geq ||\rho||^2 - (\delta(\Gamma)-\rho_{min})^2$ (cf. Theorem 3.15), we obtain a slight improvement of Corollary 3.11:

Corollary 3.24. *Let $\rho_{min} \leq \delta(\Gamma) < \rho_{min} + ||\rho||$. Then there is for all $\varepsilon > 0$ and $\mu \in (0,1)$ a constant $C_{\varepsilon,\mu}$, such that*

$$\begin{aligned} k(t,\tilde{x},\tilde{y}) \;\; &\leq \;\; C_{\varepsilon,\mu} t^{-n/2} \exp\left((2\mu - (||\rho||^2 - (\delta(\Gamma)-\rho_{min})^2))t - \frac{d_M^2(\tilde{x},\tilde{y})}{4(1+\mu)t} \right) \\ &\cdot \sqrt{P(\delta(\Gamma)+\varepsilon;\tilde{x},\tilde{x})}\sqrt{P(\delta(\Gamma)+\varepsilon;\tilde{y},\tilde{y})}. \end{aligned}$$

For the remaining part of this section we assume that $X = G/K$ is a symmetric space of non-compact type. Note, that the Lie group G is therefore semi-simple. If we assume that G possesses moreover Kazhdan's property (T), we can find a constant $\tilde{c}(G) > 0$ (only depending on G), such that for all locally symmetric spaces $M = \Gamma\backslash X$ the following holds (cf.[56]):

$$||\rho||^2 \geq \lambda_0(M) \geq \tilde{c}(G) > 0.$$

Together with Corollary 3.22 we obtain

Corollary 3.25. *Assume $X = G/K$ is a symmetric space of non-compact type and G has Kazhdan's property (T). Then there are universal constants $c(G), \mu(G) > 0$, such that the following estimate holds for the heat kernel on all locally symmetric spaces $\Gamma\backslash X$:*

$$\begin{aligned} k(t,\tilde{x},\tilde{y}) \;&\leq\; C_\varepsilon t^{-n/2} \exp\left(-c(G)t - \frac{d_M^2(\tilde{x},\tilde{y})}{4(1+\mu(G))t}\right) \cdot \\ &\cdot \sqrt{P(\delta(\Gamma)+\varepsilon,\tilde{x},\tilde{x})}\sqrt{P(\delta(\Gamma)+\varepsilon,\tilde{y},\tilde{y})}, \end{aligned}$$

for all $\varepsilon > 0$.

Proof. We define $\mu(G) := \min\{\frac{1}{2}, \frac{c'(G)}{4}\} \in (0,1)$ and $c(G) := -2\mu(G) + \tilde{c}(G) > 0$. □

3.6. Lower Bounds

In this section we determine lower bounds for the heat kernel k on locally symmetric spaces $\Gamma\backslash X$. Note, that these bounds hold only pointwise on the diagonal.

Lemma 3.26. *Let M denote an arbitrary Riemannian manifold. Then, for any $\varepsilon > 0$ and $x \in M$, there is a constant $C_x > 0$ such that for all $t > 0$ the estimate*

$$K(t,x,x) \geq C_x \exp\left(-(\lambda_0(M)+\varepsilon)t\right)$$

holds.

Proof. We choose a pre-compact subset $\Omega \subset\subset M$ with $x \in \Omega$ and $\lambda_0(\Omega) \leq \lambda_0(M) + \varepsilon$. We conclude

$$K(t,x,x) \geq K_\Omega(t,x,x) = \sum_{j=0}^{\infty} \mathrm{e}^{-\lambda_j(\Omega)t}\Phi_{j,\Omega}^2(x) \geq \mathrm{e}^{-\lambda_0(\Omega)t}\Phi_{0,\Omega}^2(x).$$

As $\Phi_{0,\Omega}(x) \neq 0$ (cf. [70], Lemma VI.3.10) the claim follows. □

Together with Theorem 3.18 we obtain the next corollary.

Corollary 3.27. *Assume $M := \Gamma\backslash X$ is a locally symmetric space. Then we have the following lower bounds for the heat kernel k on M:*

(a) *Let $0 \leq \delta(\Gamma) \leq ||\rho||$. For any $\varepsilon > 0$ and $\tilde{x} \in M$ there is a constant $C_{\tilde{x}}$ such that the estimate*

$$k(t,\tilde{x},\tilde{x}) \geq C_{\tilde{x}} \exp\left(-(||\rho||^2 + \varepsilon)t\right)$$

holds.

(b) *If* $||\rho|| \leq \delta(\Gamma) \leq 2||\rho||$, *there is for any* $\varepsilon > 0$ *and* $\tilde{x} \in M$ *a constant* $C_{\tilde{x}}$ *such that the estimate*

$$k(t, \tilde{x}, \tilde{x}) \geq C_{\tilde{x}} \exp\left(-(||\rho||^2 - (\delta(\Gamma) - ||\rho||)^2 + \varepsilon)t\right)$$

holds.

Let us compare these lower bounds with the upper bounds obtained in Theorem 3.6 and in Theorem 3.7:
For subgroups Γ with $\delta(\Gamma) < \rho_{min}$ in both cases the term $e^{-||\rho||^2 t}$ is dominant if one is interested in the asymptotic behavior for $t \to \infty$ (pointwise on the diagonal).
If $\delta(\Gamma) \geq \rho_{min}$, the lower bound and the upper bound differ more and more the larger the difference $||\rho|| - \rho_{min}$ becomes. Since for rank-1 symmetric spaces X we have $||\rho|| = \rho_{min}$, the dominant terms (for $t \to \infty$) coincide also in the rank-1 case for subgroups with $\delta(\Gamma) \geq ||\rho||$.

3.7. L^2-Eigenfunctions of the Laplace-Beltrami operator

We apply some of the previous results in this chapter to prove that every L^2-eigenfunction of the Laplace-Beltrami operator on a locally symmetric space with bounded geometry is bounded. Note that a locally symmetric space $M = \Gamma\backslash X$ has *bounded geometry* if and only if the injectivity radius $\mathrm{inj}(M)$ of M is bounded away from zero. If M has bounded geometry and in addition finite volume, the manifold M is necessarily compact since no cusps can occur. Therefore the theorem below is interesting only if the volume of M is infinite.

Theorem 3.28. *Let* $M := \Gamma\backslash X$ *denote a locally symmetric space with bounded geometry and* $\dim M \geq 3$. *Then every eigenfunction* $\varphi \in L^2(M)$ *of* Δ_M *is bounded.*

The proof of Theorem 3.28 is an easy consequence of the next lemma.

Lemma 3.29. *Assume, the locally symmetric space* $M := \Gamma\backslash X$ *has bounded geometry and* $n := \dim M \geq 3$. *Then there exists a positive function* $t \mapsto C(t)$ *such that*

$$||e^{-t\Delta_M} f||_{L^\infty} \leq C(t)||f||_{L^2}$$

for all $f \in L^2(M)$. *More precisely, we can choose*

$$C(t) = ct^{-n/4}e^{\mu t}$$

for certain positive constants c *and* μ.

Proof. We first remind the reader of the heat kernel estimate (cf. Corollary 3.22)

$$k(t,\tilde{x},\tilde{y}) \leq C_1 t^{-n/2} \mathrm{e}^{2\mu t} \sqrt{P(s;\tilde{x},\tilde{x})}\sqrt{P(s;\tilde{y},\tilde{y})},$$

where $s > \delta(\Gamma)$. If we choose $s > 2||\rho||$, the Poincaré series is (for fixed s) bounded (cf. Corollary 3.4). We obtain

$$k(t,\tilde{x},\tilde{y}) \leq C_2 t^{-n/2} \mathrm{e}^{2\mu t} =: \tilde{C}(t).$$

This yields for any $f \in L^1(M)$ the estimate

$$||\mathrm{e}^{-t\Delta_M} f||_{L^\infty} \leq \tilde{C}(t)||f||_{L^1}.$$

Therefore, $\mathrm{e}^{-t\Delta_M} : L^1(M) \to L^\infty(M)$ is bounded and the operator norm is bounded by $\tilde{C}(t)$.

Because of $\int_M k(t,\tilde{x},\tilde{y})\, dvol(\tilde{y}) \leq 1$, we have for all $f \in L^\infty(M)$:

$$||\mathrm{e}^{-t\Delta_M} f||_{L^\infty} \leq ||f||_{L^\infty}.$$

Applying the Riesz-Thorin interpolation theorem we obtain the boundedness of the operators $\mathrm{e}^{-t\Delta_M} : L^2(M) \to L^\infty(M)$ and further

$$||\mathrm{e}^{-t\Delta_M}||_{L^2 \to L^\infty} \leq \tilde{C}(t)^{1/2}.$$

□

Now, we are prepared for the proof of Theorem 3.28.

Proof of Theorem 3.28. We first remark that the self-adjoint operator Δ_M is unitarily equivalent to a multiplication operator T_f by the spectral theorem for unbounded self-adjoint operators:

$$U\Delta_M U^* = T_f.$$

For the semigroup $\mathrm{e}^{-t\Delta_M}$ it follows $U\mathrm{e}^{-t\Delta_M}U^* = T_{\mathrm{e}^{-tf}}$. If the function $\varphi \in L^2(M)$ is an eigenfunction of Δ_M with eigenvalue λ, we can conclude that φ is also an eigenfunction of $\mathrm{e}^{-t\Delta_M}$ with eigenvalue $\mathrm{e}^{-t\lambda}$. We therefore obtain:

$$\mathrm{e}^{-\lambda}||\varphi||_{L^\infty} = ||\mathrm{e}^{-\lambda}\varphi||_{L^\infty} = ||\mathrm{e}^{-\Delta_M}\varphi||_{L^\infty} \leq \tilde{c}\,||\varphi||_{L^2},$$

where we used in the last step the preceding lemma. As φ lies in $L^2(M)$ the claim follows. □

The last proof implies immediately

Corollary 3.30. *Let* $M := \Gamma\backslash X$ *denote a locally symmetric space with bounded geometry and* $\dim M \geq 3$. *If* φ_λ *is an* L^2*-eigenfunction of* Δ_M *with eigenvalue* λ *and* $||\varphi||_{L^2} = 1$, *the following holds:*

$$||\varphi_\lambda||_{L^\infty} \leq \tilde{c}\,\mathrm{e}^{\lambda}.$$

Chapter 4.

Locally Symmetric Spaces with Small Fundamental Group

In [75] M. E. Taylor showed that the L^p-spectrum of the Laplace-Beltrami operator on a symmetric space of non-compact type coincides with a certain parabolic region P_p which degenerates in the case $p = 2$ to the interval $P_2 = [||\rho||^2, \infty)$.
He also proved that the L^p-spectrum of certain locally symmetric spaces $M = \Gamma\backslash X$ is – except for a finite subset of $\mathbb{R}$ – contained in the parabolic region P_p. More precisely, we have:

Proposition 4.1 (cf. [75], Proposition 3.3). *Let X denote a symmetric space of non-compact type and $M = \Gamma\backslash X$ a locally symmetric space with either finite volume or bounded injectivity radius, i.e.* $\text{inj}(M) \geq c > 0$. *If*

$$\sigma(\Delta_{M,2}) \subset \{\lambda_0, \ldots, \lambda_m\} \cup [||\rho||^2, \infty), \tag{4.1}$$

where $\lambda_j \in [0, ||\rho||^2)$ are eigenvalues of finite multiplicity, then we have for $p \in [1, \infty)$:

$$\sigma(\Delta_{M,p}) \subset \{\lambda_0, \ldots, \lambda_m\} \cup P_p.$$

However, the L^2-spectrum of locally symmetric spaces (with infinite volume) is in general unknown. Moreover, for finite volume locally symmetric spaces the assumption (4.1) is in general false (cf. Chapter 5).

Our main concern in this chapter is to prove that for certain locally symmetric spaces $M = \Gamma\backslash X$ (with *infinite* volume) the L^p-spectrum of the respective Laplace-Beltrami operator coincides with the parabolic region P_p and therefore with the L^p-spectrum of the universal cover X of M:

Theorem 4.2. *Let $X = G/K$ denote a symmetric space of non-compact type with* $\operatorname{rank}(X) = 1$ *and* $\dim(X) \geq 3$. *Assume, that the locally symmetric space $M = \Gamma\backslash X$ has bounded geometry and that Γ is small. Then we have for $p \in [1,\infty)$:*

$$\sigma(\Delta_{M,p}) = P_p = \sigma(\Delta_{X,p}).$$

The precise definitions of a small subgroup and the parabolic region P_p will be given in Definition 4.22 and Definition 4.24. In the case $p = 3$ the parabolic region P_p is indicated in Figure 4.1 as the hatched area. The dashed rays in this figure are tangent to the parabolic region P_p and are the boundary of the sector

$$\left\{ z \in \mathbb{C} \setminus \{0\} : |\arg(z)| \leq \arctan \frac{|p-2|}{2\sqrt{p-1}} \right\} \cup \{0\}$$

from Section 2.2. Since the L^p-spectrum has to be contained in the sector from above, this proves that the estimate of V.A. Liskevich and M.A. Perelmuter for the angle of analyticity θ_p for submarkovian semigroups, cf. (2.3), is optimal.

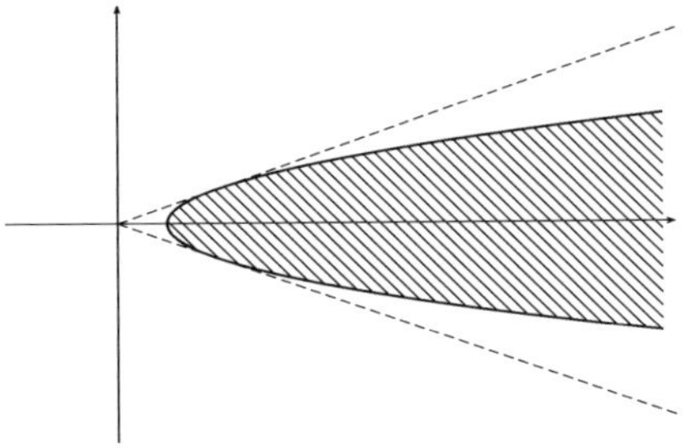

Figure 4.1.: The parabolic region P_p if $p = 3$.

Proposition 4.26 below implies that we can omit the assumption "$M = \Gamma\backslash X$ has bounded geometry" in the case $p = 2$. Thus, the L^2-spectrum in this situation is given by

$$\sigma(\Delta_{M,2}) = [||\rho||^2, \infty) = \sigma(\Delta_{X,2})$$

and coincides with the L^2-spectrum of the universal covering space X. This generalizes a result due to K. Corlette (cf. Theorem 4.2 in [16] or Theorem 3.18) who proved that the bottom of the L^2-spectrum λ_0 is in both cases given by $||\rho||^2$.

Usually, one proves such a result by the use of a Fourier type transform that turns the Laplace-Beltrami operator into a multiplication operator. In the universal covering space X for example, we have the Helgason-Fourier transform (cf. [42]), and for arithmetic non-uniform lattices Γ the theory of Eisenstein series can be used in order to determine the L^2-spectrum of the Laplace-Beltrami operator on $\Gamma\backslash X$ (cf. Chapter 5 and the references given there). In the general case however, we have no Fourier transform.

4.1. Preliminaries

4.1.1. Horocyclic Coordinates

Let $X = G/K$ denote a rank-1 symmetric space of non-compact type. Recall, that in this case $G := \mathrm{Isom}^0(X)$ is a semi-simple Lie group. We further assume that the Riemannian metric $\langle\cdot,\cdot\rangle$ of X in eK coincides with the restriction of the Killing form B to $\mathfrak{p} \cong T_{eK}G/K$, which is not a great restriction (cf. Section 1.2).

We denote by $\mathfrak{g}$ the Lie algebra of G, by $\mathfrak{g} = \mathfrak{k} \oplus \mathfrak{p}$ the Cartan decomposition, and by $\mathfrak{a} \subset \mathfrak{p}$ a maximal abelian subalgebra.

The choice of a Weyl chamber $\mathfrak{a}^+$ determines an Iwasawa decomposition $G = NAK$. This yields a diffeomorphism

$$\mu : N \times A \to X,\ (n, a) \mapsto na \cdot x_0.$$

We remind the reader of the fact $N = \exp\left(\sum_{\beta\in\Sigma^+} \mathfrak{g}_\beta\right)$, where Σ^+ denotes the set of positive (restricted) roots of the pair $(\mathfrak{g}, \mathfrak{a})$ with respect to the Weyl chamber $\mathfrak{a}^+$. Note also, that the root spaces $\mathfrak{g}_\alpha$ and $\mathfrak{g}_\beta$ for $\alpha \neq \beta \in \Sigma^+$ are orthogonal with respect to the scalar product $\langle\cdot,\cdot\rangle$. We choose for each $\alpha \in \Sigma^+$ an orthonormal basis of $\mathfrak{g}_\alpha$ and the union $\{N_1, \ldots, N_{n-1}\}$ of these bases yields an orthonormal basis of the subspace $\mathfrak{n}$. Furthermore, we take a unit vector H from the 1-dimensional subalgebra $\mathfrak{a}$. This leads to a global coordinate map

$$\begin{aligned} \varphi \ &: \ X = NA \cdot x_0 \to \mathbb{R}^n, \\ &\exp\left(\sum_{i=1}^{n-1} x_i N_i\right) \exp(yH) \cdot x_0 \mapsto (x_1, \ldots, x_{n-1}, y) =: (x, y). \end{aligned}$$

We call these coordinates *horocyclic coordinates* since the orbits $N \cdot x$ are usually called horocycles in X. In the rank-1 case, the orbits under N are horospheres but for higher rank symmetric spaces this is not true as horospheres always have codimension one (for more details cf. [47]).

4.1.2. The Metric in Horocyclic Coordinates

For all $\alpha \in \Sigma^+$ we define a left invariant bilinear form h_α on the nilpotent subgroup $N = \exp\left(\sum_{\beta\in\Sigma^+} \mathfrak{g}_\beta\right)$ by

$$h_\alpha := \begin{cases} \langle\cdot,\cdot\rangle, & \text{on } \mathfrak{g}_\alpha, \\ 0, & \text{else,} \end{cases}$$

where $\langle\cdot,\cdot\rangle$ denotes again the scalar product defined by means of the Killing form. Then we have the following formula for the pullback μ^*g of the metric g on X to $N \times A$.

Proposition 4.3. *Denoting by* $\log a$ *the unique* $H \in \mathfrak{a}$ *with* $\exp H = a$*, the pullback* $\mu^* g$ *is given by*

$$ds^2_{(n,a)} = \left(\frac{1}{2} \sum_{\alpha \in \Sigma^+} \mathrm{e}^{-2\alpha(\log a)} h_\alpha \right) \oplus da^2.$$

Proof. The proof follows directly from Proposition 1.6 in [6] or Proposition 4.3 in [7] respectively. □

We immediately obtain the following three important corollaries.

Corollary 4.4. *The representation of the metric* g *with respect to horocyclic coordinates is given by*

$$(g_{ij})_{i,j}(na \cdot x_0) = \left(\begin{array}{ccc|c} 2^{-1}\mathrm{e}^{-2\alpha_1(\log a)} & & & \\ & \ddots & & 0 \\ & & 2^{-1}\mathrm{e}^{-2\alpha_{n-1}(\log a)} & \\ \hline & 0 & & 1 \end{array} \right),$$

where the roots $\alpha_i \in \Sigma^+$ *appear in the matrix above according to their multiplicity and the indices are chosen in a way such that*

$$\left(\sum_{\alpha \in \Sigma^+} \mathrm{e}^{-2\alpha(\log a)} h_\alpha \right) (N_i, N_i) = \mathrm{e}^{-2\alpha_i(\log a)}.$$

Corollary 4.5. *Let* X *denote a rank-*1 *symmetric space of non-compact type. The volume form of* X *with repect to horocyclic coordinates is*

$$\begin{aligned} \sqrt{\det(g_{ij})(na \cdot x_0)}\, dxdy &= \left(\frac{1}{2}\right)^{(n-1)/2} \mathrm{e}^{-2\rho(\log a)}\, dxdy \\ &= \left(\frac{1}{2}\right)^{(n-1)/2} \mathrm{e}^{-2y\rho(H)}\, dxdy, \end{aligned}$$

where $\log a = yH$.

Corollary 4.6. *Let* X *denote a rank-*1 *symmetric space of non-compact type. If we choose in the definition of horocyclic coordinates* $H \in \mathfrak{a}^+$ *with* $||H|| = 1$*, the Laplace-Beltrami operator in these coordinates is given by*

$$\Delta_X = -2 \sum_{i=1}^{n-1} \mathrm{e}^{2\alpha_i} \frac{\partial^2}{\partial x_i^2} + 2||\rho|| \frac{\partial}{\partial y} - \frac{\partial^2}{\partial y^2},$$

where $\mathrm{e}^{2\alpha_i}$ *is short hand for the function* $(x, y) \mapsto \mathrm{e}^{2y\alpha_i(H)}$.

Proof. Remember the formula

$$\Delta_X = -\frac{1}{\det(g_{ij})^{1/2}} \sum_{i,j} \partial_i(g^{ij} \det(g_{ij})^{1/2} \partial_j)$$

and notice that $\rho(H) = ||\rho||$. Then a straightforward calculation shows the assertion. □

4.1.3. Geodesic Compactification and Limit sets

Let X denote in this subsection always a symmetric space of non-compact type. Our main references are [29, 30].

Geodesic Compactification

Definition 4.7. *Two unit speed geodesics c_1 and c_2 of X are called* asymptotic *if there exists a positive constant C such that*

$$d(c_1(t), c_2(t)) \leq C \qquad \textit{for all } \ t \geq 0.$$

This clearly defines an equivalence relation on the unit speed geodesics of X. The set of equivalence classes is called the geodesic boundary *of X and denoted by $X(\infty)$. We further put*

$$\overline{X} := X \cup X(\infty).$$

If c is a unit speed geodesic in X, the respective equivalence class is denoted by $c(\infty)$.

For the proof of the next proposition we refer to Proposition 1.2 in [30].

Proposition 4.8. *Let c_0 denote a unit speed geodesic in X. Then for any point $x \in X$ there exists a unique unit speed geodesic c in the equivalence class of c_0 with $c(0) = x$.*

Because of this proposition, the geodesic boundary $X(\infty)$ can be identified with the unit sphere $S_x X$ in any tangent space $T_x X$, in particular with the unit sphere in $\mathfrak{p} \cong T_{x_0} X$.

Example. If we represent the hyperbolic plane $\mathbb{H}^2$ as the open unit disc in $\mathbb{R}^2$ with metric

$$ds^2 = \frac{4}{(1 - x^2 - y^2)^2}(dx^2 + dy^2),$$

the unparameterized geodesics are the Euclidean circles orthogonal to the unit circle S^1, and the diameters of the unit disc. Two geodesics are asymptotic if and only if they meet S^1 in the same point. Therefore, $\mathbb{H}^2(\infty) \cong S^1$.

Notation. (1) By c_{xy} we denote the unit speed geodesic connecting x and y with $c_{xy}(0) = x$.
(2) For $v \in S_xX$ we denote by c_v the unit speed geodesic with $c_v(0) = x$ and $c_v'(0) = v$.
(3) For $x \in X$ and $\xi \in X(\infty)$ we denote by $c_{x\xi}$ the unique unit speed geodesic in the equivalence class ξ with $c_{x\xi}(0) = x$.
(4) Let c denote a unit speed geodesic with $c(0) = x_0$. For any $\varepsilon > 0$ and $r > 0$ we define the *truncated cone* (with vertex x_0, axis c, and angle ε) by

$$C(c, \varepsilon, r) := \{x \in \overline{X} : \sphericalangle_{x_0}(c'(0), c'_{x_0x}(0)) < \varepsilon\} \setminus \{x \in X : d(x_0, x) \leq r\}.$$

Proposition 4.9 (cf. Proposition 2.3 in [30]). *There is a unique topology τ on $\overline{X} = X \cup X(\infty)$ such that:*

1. *Given $x \in X$ and $\xi \in X(\infty)$ the truncated cones $C(c_{x\xi}, \varepsilon, r)$ with $\varepsilon > 0, r > 0$ form a neighborhood basis for the topology at ξ.*
2. *The topology of X induced from τ coincides with the original topology of X and X is a dense open subset of $\overline{X}$.*

This topology τ is called cone topology.

The following theorem indicates an analogy with the Poincaré ball model for the hyperbolic space.

Theorem 4.10 (cf. Theorem 2.10 in [30]). *Choose $x \in X$ and denote by $B(x)$ the closed unit ball with boundary sphere S_xX in the tangent space T_xX. Let $f : [0, 1] \to [0, \infty]$ be a homeomorphism. Then the map*

$$\varphi : B(x) \to \overline{X},\ v \mapsto \exp\left(f(||v||)v\right)$$

is a homeomorphism with $\varphi(S_xX) = X(\infty)$.

Since asymptotic geodesics are mapped onto asymptotic geodesics by isometries, we also have the following result.

Lemma 4.11. *Any isometry of X extends to a homeomorphism of $\overline{X}$.*

Limit Sets

Definition 4.12. *Let $\Gamma \subset \mathrm{Isom}(X)$ denote an arbitrary subgroup of the isometry group. The* limit set $\Lambda(\Gamma)$ *of Γ is defined as*

$$\Lambda(\Gamma) := \overline{\Gamma \cdot x} \cap X(\infty),$$

where $\overline{\Gamma \cdot x}$ is the closure of the orbit $\Gamma \cdot x$ in $\overline{X}$.

Note, that the limit set $\Lambda(\Gamma)$ does not depend on the choice of $x \in X$ (cf. [29], p.33).

Lemma 4.13. (1) *The limit set* $\Lambda(\Gamma)$ *is a closed subset of* $X(\infty)$.

(2) *If* $\Gamma \subset \mathrm{Isom}(X)$, $\Gamma \neq \{\mathrm{id}\}$, *is a discrete, freely acting subgroup, the limit set* $\Lambda(\Gamma)$ *is always non-empty.*

Proof. The first assertion follows directly from the definition. For a proof of the second assertion we refer to [29], p.33. □

In accordance with the theory of discrete subgroups in hyperbolic geometry we make the following definition (see e.g. [67], p.577).

Definition 4.14. *A discrete, freely acting subgroup* $\Gamma \subset \mathrm{Isom}(X)$ *is called a subgroup of the* first kind *if* $\Lambda(\Gamma) = X(\infty)$ *and of the* second kind *otherwise.*

For a subgroup of the second kind, the *ordinary set*

$$\Omega(\Gamma) := X(\infty) \setminus \Lambda(\Gamma)$$

is always a non-empty open subset of the geodesic boundary $X(\infty)$.

If $\mathrm{rank}(X) = 1$, $\Omega(\Gamma)$ is often called *region of discontinuity.* The following proposition justifies this.

Proposition 4.15 (cf. Proposition 8.5 in [30]). *Let X denote a symmetric space of non-compact type with* $\mathrm{rank}(X) = 1$ *and* $\Gamma \subset \mathrm{Isom}(X)$ *a discrete, freely acting subgroup of isometries. Then the set of points of* $\overline{X}$ *at which the action of* Γ *is properly discontinuous is* $X \cup \Omega(\Gamma)$.

Example 4.16. Let $X = G/K$ denote a rank one symmetric space of non-compact type and $G = NAK$ an Iwasawa decomposition of the semi-simple Lie group $G := \mathrm{Isom}^0(X)$. A discrete subgroup $\Gamma = \langle a \rangle$, $a \neq \mathrm{id}$ of the one dimensional Lie group A is always a second kind subgroup. This can be seen as follows:
If $x_0 \in X$ denotes the basepoint for the Iwasawa decomposition and if $H = \log(a) \in \mathfrak{a}$ is the unique element in the Lie algebra of A with $\exp(H) = a$, the unparameterized geodesic $A{\cdot}x_0$ is given by the image of the curve $c(t) = \exp(tH){\cdot}x_0$. The equivalence class of the geodesic $c(t)$ (resp. $c(-t)$) is denoted by $c(\infty)$ (resp. $c(-\infty)$). Hence, the limit set $\Lambda(\Gamma)$ consists only of the two points $c(\infty)$ and $c(-\infty) \in X(\infty)$.

Since we want to work with Dirichlet fundamental domains in the next section, we present some results here.

Recall, that for a discrete, freely acting subgroup of isometries Γ and $x_0 \in X$ the set

$$F = F(x_0) := \{x \in X : d(x, x_0) < d(\gamma x, x_0) \text{ for all } \gamma \in \Gamma \setminus \{\mathrm{id}\}\}$$

is called *Dirichlet fundamental domain* (with respect to $x_0 \in X$). An easy calculation shows that F is star-shaped with respect to x_0.

Lemma 4.17 (cf. Corollary 3.3.2 in [11]). *Let X denote a symmetric space of non-compact type with* $\operatorname{rank}(X) = 1$ *and* $\Gamma \subset \operatorname{Isom}(X)$ *a discrete, freely acting subgroup of isometries. If* $F = F(x_0)$ *is a Dirichlet fundamental domain, the set*

$$\Gamma \cdot F$$

is locally finite on $X \cup \Omega(\Gamma)$.

The following two theorems have their analogues in hyperbolic geometry (see e.g. Theorem 12.1.13 and Theorem 12.1.15 in [67]). For the lack of knowledge of a reference in the general rank 1 case, we give proofs here.

Theorem 4.18. *Let X denote a symmetric space of non-compact type with* $\operatorname{rank}(X) = 1$ *and* $F = F(x_0)$ *a Dirichlet fundamental domain for a discrete, freely acting subgroup* $\Gamma \subset \operatorname{Isom}(X)$ *of isometries. Then*

$$\Omega(\Gamma) = \bigcup_{\gamma \in \Gamma} \gamma \left(\overline{F} \cap \Omega(\Gamma) \right),$$

where $\overline{F}$ denotes the closure of F in $\overline{X}$.

Proof. Since the ordinary set $\Omega(\Gamma)$ is Γ-invariant, it is clear that the right hand side is contained in the left hand side. In order to prove the other inclusion, we choose $\xi \in \Omega(\Gamma)$. Then there exists a sequence $(x_i)_{i \in \mathbb{N}}$ in X converging to ξ. For any $i \in \mathbb{N}$ we can find an isometry $\gamma_i \in \Gamma$ such that $x_i \in \gamma_i F$. Since $\Gamma \cdot F$ is locally finite on $X \cup \Omega(\Gamma)$, only finitely many isometries of the sequence $(\gamma_i)_i$ are distinct. Hence, there is a $j \in \mathbb{N}$ such that

$$x_i \in \gamma_j F$$

for infinitely many $i \in \mathbb{N}$. Therefore, we have $\xi \in \gamma_j \overline{F}$ and in conclusion

$$\xi \in \gamma_j \overline{F} \cap \Omega(\Gamma) = \gamma_j \left(\overline{F} \cap \Omega(\Gamma) \right).$$

□

Theorem 4.19. *Let X denote a symmetric space of non-compact type with* $\operatorname{rank}(X) = 1$ *and* $\Gamma \subset \operatorname{Isom}^0(X)$ *a subgroup of the second kind. Then*

$$\operatorname{vol}(\Gamma \backslash X) = \infty.$$

Proof. Let $F = F(x_0)$ be a Dirichlet fundamental domain for Γ. It clearly suffices to prove $\operatorname{vol}(F) = \infty$.

The ordinary set $\Omega(\Gamma)$ is non-empty and homeomorphic to an open subset of the unit sphere in any tangent space. In particular, the ordinary set is a Baire space. As a discrete subgroup, Γ is countable and therefore one of the closed subsets $\gamma\left(\overline{F}\cap\Omega(\Gamma)\right)$ of $\Omega(\Gamma)$ must have non-empty interior. Thus, the interior of $\overline{F}\cap\Omega(\Gamma)$ is non-empty.

Choose an interior point ξ of $\overline{F}\cap\Omega(\Gamma)$. Since F is star-shaped with respect to x_0, the geodesic ray $c_{x_0\xi}:[0,\infty)\to X$, $t\mapsto c_{x_0\xi}(t)$ is completely contained in F. If $U\subset\overline{F}\cap\Omega(\Gamma)$ is an open neighborhood of ξ, every geodesic ray starting at x_0 in the equivalence class of some $\eta\in U$ is contained in F for the same reason. Hence, there are $\varepsilon>0$ and $r>0$ such that the truncated cone $C(c_{x_0\xi},\varepsilon,r)$ is contained in $\overline{F}$. Let us take an Iwasawa decomposition

$$G\cong N\times A\times K$$

of $G=\mathrm{Isom}^0(X)$ where K is the isotropy subgroup of G with respect to x_0, $A\cdot x_0=c_{x_0\xi}(\mathbb{R})$, and if $H\in\mathfrak{a}$ (the Lie algebra of A) is defined by the equation

$$\exp(tH)\cdot x_0=c_{x_0\xi}(-t),$$

the positive Weyl chamber of $\mathfrak{a}$ is

$$\mathfrak{a}^+:=\{tH:t>0\}.$$

Furthermore, we choose a compact neighborhood N_0 of the identity in N such that for all $n_0\in N_0$ and $t\geq r+1$ the geodesic

$$t\mapsto n_0\cdot c_{x_0\xi}(t)$$

is contained in the truncated cone $C(c_{x_0\xi},\varepsilon,r)$. We claim that the "box"

$$N_0\cdot c_{x_0\xi}(\mathbb{R}_{\geq r+1})$$

has infinite volume. Since this box is completely contained in the fundamental domain $F(x_0)$, this proves

$$\mathrm{vol}\,(\Gamma\backslash X)=\mathrm{vol}\,(F)=\infty.$$

We use horocyclic coordinates (cf. Section 4.1.1 and 4.1.2) and obtain

$$\begin{aligned}\mathrm{vol}\,(N_0\cdot c_{x_0\xi}(\mathbb{R}_{\geq r+1})) &= \left(\frac{1}{2}\right)^{(n-1)/2}\int_{\varphi(N_0\cdot c_{x_0\xi}(\mathbb{R}_{\geq r+1}))} \mathrm{e}^{-2y\rho(H)}\,dxdy\\ &= \left(\frac{1}{2}\right)^{(n-1)/2}\int_{\varphi(N_0\cdot x_0)}dx\int_{-\infty}^{-r-1}\mathrm{e}^{-2y\|\rho\|}\,dy\\ &= \infty.\end{aligned}$$

□

Definition 4.20. *Let X denote a symmetric space of non-compact type with* $\mathrm{rank}(X) = 1$. *A discrete subgroup $\Gamma \subset \mathrm{Isom}^0(X)$ that acts without fixed points on X is called* geometrically cocompact *if the Kleinian manifold*

$$\overline{M}_\Gamma := (X \cup \Omega(\Gamma))\,/\Gamma$$

is compact.

In [16], Theorem 5.2, K. Corlette proved the following result.

Theorem 4.21. *Let X denote a symmetric space of non-compact type with* $\mathrm{rank}(X) = 1$ *and $\Gamma \subset \mathrm{Isom}^0(X)$ a geometrically cocompact subgroup. Then, the Hausdorff dimension of the limit set $\Lambda(\Gamma)$ is bounded from above by the critical exponent $\delta(\Gamma)$.*

We can conclude for geometrically cocompact subgroups Γ with $\delta(\Gamma) \leq ||\rho||$ that the (closed) limit set $\Lambda(\Gamma)$ does not have full Hausdorff dimension. Consequently, Γ is a subgroup of the second kind.

Definition 4.22. *Let X be a symmetric space of non-compact type with* $\mathrm{rank}(X) = 1$. *A subgroup $\Gamma \subset \mathrm{Isom}^0(X)$ of the second kind is called* small *if $\delta(\Gamma) \leq ||\rho||$, where $\delta(\Gamma)$ denotes the critical exponent of Γ.*

Example 4.23. The subgroup $\Gamma = \langle a \rangle$ in Example 4.16 is small. To see this, it only remains to check whether its critical exponent is small. Recall, that the critical exponent of Γ can be defined as follows:

$$\delta(\Gamma) := \inf\{s > 0 : \sum_{n \in \mathbb{Z}} \mathrm{e}^{-sd(x_0, a^n x_0)} < \infty\},$$

where $x_0 \in X$ denotes the basepoint that corresponds to the maximal compact subgroup K. If $H \in \mathfrak{a}$ is the unique element in the Lie algebra $\mathfrak{a}$ of A such that $a = \exp H$, we obtain

$$d(x_0, a^n x_0) = d(x_0, \exp(nH)x_0) = |n| \cdot ||H||$$

and therefore for any $s > 0$

$$\sum_{n \in \mathbb{Z}} \mathrm{e}^{-sd(x_0, a^n x_0)} = \sum_{n \in \mathbb{Z}} \mathrm{e}^{-s|n| \cdot ||H||} = \frac{1 + \mathrm{e}^{-s||H||}}{1 - \mathrm{e}^{-s||H||}} < \infty.$$

In particular, we may conclude $\delta(\Gamma) = 0$.

4.2. L^p-Spectrum

In this section we denote by $X = G/K$ a symmetric space of non-compact type with $\operatorname{rank}(X) = 1$. We further assume as above that the restriction of the metric to the tangent space $T_{x_0}X \cong \mathfrak{p}$ of the basepoint x_0 equals the restriction of the Killing form to $\mathfrak{p}$.

Definition 4.24. *For $p \in [1, \infty)$ we put*

$$P_p := \left\{ z = x + iy \in \mathbb{C} : x \geq \frac{4||\rho||^2}{p}\left(1 - \frac{1}{p}\right) + \frac{y^2}{4||\rho||^2(1-\frac{2}{p})^2} \right\}$$

if $p \neq 2$ and $P_2 := [||\rho||^2, \infty)$.

Notice, that the boundary ∂P_p of the parabolic region P_p is given by the curve

$$\begin{aligned} \mathbb{R} \to \mathbb{C}, \quad s \mapsto \quad & \frac{4||\rho||^2}{p}\left(1 - \frac{1}{p}\right) + s^2 + 2i||\rho||s\left(1 - \frac{2}{p}\right) \\ &= \left(\frac{2||\rho||}{p} + is\right)\left(2||\rho|| - \frac{2||\rho||}{p} - is\right). \end{aligned}$$

One part of the proof of Theorem 4.2 is a consequence of Proposition 3.3 in [75] (cf. Proposition 4.1). The constraint on the dimension of X and that the subgroup Γ is of the second kind will not be important in this part:

Proposition 4.25. *Let $X = G/K$ denote a symmetric space of non-compact type with* $\operatorname{rank}(X) = 1$, *$\Gamma \subset G$ a discrete, freely acting subgroup of G such that $M = \Gamma\backslash X$ has bounded geometry and $\delta(\Gamma) \leq ||\rho||$. Then we have for $p \in [1, \infty)$:*

$$\sigma(\Delta_{M,p}) \subset P_p.$$

Proof. Recall, that

$$\sigma(\Delta_{M,2}) \subset [||\rho||^2, \infty)$$

(cf. Theorem 3.18), and that we have the following estimate for the volume of a ball $B(r) \subset X$ with radius $r > 0$:

$$\operatorname{vol} B(r) \leq C e^{2||\rho||r}$$

(cf. [50]). In view of Proposition 3.3 in [75] this is enough to prove the assertion. □

The following proposition is a crucial step towards the proof of the reverse inclusion in Theorem 4.2.

Proposition 4.26. *Let $X = G/K$ denote a symmetric space of non-compact type with* $\operatorname{rank}(X) = 1$, *$\Gamma \subset G$ a subgroup of the second kind, and $M = \Gamma\backslash X$ the respective locally symmetric space. Then, for any $p \in [1, \infty)$, the boundary ∂P_p of the parabolic region P_p is contained in the approximate point spectrum of $\Delta_{M,p}$:*

$$\partial P_p \subset \sigma_{app}(\Delta_{M,p}).$$

Remark 4.27. We know that $e^{-t\Delta_{M,1}} : L^1(M) \to L^1(M)$ is an analytic semigroup if the manifold M has bounded geometry (cf. [77, 18]). A consequence of Proposition 4.26 is that the semigroup $e^{-t\Delta_{M,1}}$ for M as in Theorem 4.2 is *not* a *bounded* analytic semigroup – in contrast to the semigroups $e^{-t\Delta_{M,p}}$ for $p \in (1, \infty)$.

Since compact perturbations of any complete non-compact Riemannian manifold M do not change the continuous L^2-spectrum of Δ_M if considered as a subset of $\mathbb{R}$, the continuous spectrum depends only on the geometry at infinity of M. This follows from the so-called *decomposition principle* (cf. [28]): If $M_0 \subset M$ is a compact manifold with boundary of the same dimension as M and Δ' the Laplacian on $M \setminus M_0$ (with Dirichlet boundary conditions), then Δ_M and Δ' have the same continuous spectrum. That is why the supports $\mathrm{supp}(f_n)$ of our sequence f_n in the proof of Proposition 4.26 will leave every compact subset of M at some time.

Proof of Proposition 4.26. To prove this proposition, we will construct for each $z \in \partial P_p$ a sequence of differentiable functions $f_n \in L^p(X)$ with support in an open Dirichlet fundamental domain for Γ such that

$$\frac{||\Delta_{X,p} f_n - z f_n||_{L^p}}{||f_n||_{L^p}} \to 0 \quad \text{if} \quad n \to \infty. \tag{4.2}$$

Such a sequence (f_n) descends to a sequence of differentiable functions in $L^p(M)$ and consequently, we are done as soon as we have shown (4.2). Our proof is subdivided into two steps. First, we construct a box completely contained in a Dirichlet fundamental domain for Γ, that can be easily described by using horocyclic coordinates. This box will be the same as the box constructed in the proof of Theorem 4.19 but we need a finer description here. Secondly, we use horocyclic coordinates to construct for any $z \in \partial P_p$ a sequence of differentiable functions f_n with support in the constructed box satisfying (4.2).

We choose the basepoint $x_0 \in X$ corresponding to the maximal compact subgroup K of G. By $F = F(x_0)$ we denote the Dirichlet fundamental domain with respect to x_0 and by $\overline{F}$ we denote the closure of F in $\overline{X} = X \cup X(\infty)$. Since Γ is a second kind subgroup, the region of discontinuity $\Omega(\Gamma)$ is a non-empty open subset of $X(\infty)$ and consequently, we can find an interior point ξ in $\overline{F} \cap X(\infty)$. Let us denote by $c_{x_0\xi}$ the unique unit-speed geodesic in the equivalence class ξ with $c_{x_0\xi}(0) = x_0$. Note, that the ray $c_{x_0\xi}(\mathbb{R}_{\geq 0})$ is contained in F as the Dirichlet fundamental domain F is star-shaped with respect to its center x_0. Since the truncated cones

$$C(c_{x_0\xi}, \varepsilon, r) := \{x \in \overline{X} : \sphericalangle_{x_0}(\xi, x) < \varepsilon\} \setminus \{x \in X : d(x, x_0) \leq r\}$$

form a neighborhood basis for the cone topology at $\xi \in X(\infty)$, we can find $\varepsilon > 0$ and $r > 0$ such that

$$C(c_{x_0\xi}, \varepsilon, r) \cap X \subset F.$$

If we choose the maximal abelian subgroup $A := \{\exp(tH) : t \in \mathbb{R}\}$ in G such that $c_{x_0\xi}(t) = \exp(-tH) \cdot x_0$, and the positive Weyl chamber $\mathfrak{a}^+ = \{tH : t > 0\}$ of the one dimensional Lie algebra $\mathfrak{a}$ of A, we obtain an Iwasawa decomposition

$$G \cong N \times A \times K,$$

with a nilpotent subgroup N of G. Note, that we have $||H|| = 1$. We finally choose a neighborhood $N_0 \subset N$ of the identity such that for any $n_0 \in N_0$ the geodesic

$$t \mapsto n_0 \exp(-tH) \cdot x_0$$

is contained in the truncated cone $C(c_{x_0\xi}, \varepsilon, r)$ if $t \geq r + 1$. Hence, the box

$$N_0\{\exp(tH) : t \leq -(r+1)\} \cdot x_0$$

is contained in the fundamental domain F.

We proceed with the second step. Choose

$$z = z(s) = \left(\frac{2||\rho||}{p} + is\right)\left(2||\rho|| - \frac{2||\rho||}{p} - is\right) \in \partial P_p.$$

With respect to *horocyclic coordinates* (cf. Section 4.1.1)

$$\begin{aligned} \varphi \; : \; & X = NA \cdot x_0 \to \mathbb{R}^n, \\ & \exp\left(\sum_{i=1}^{n-1} x_i N_i\right) \exp(yH) \cdot x_0 \mapsto (x_1, \ldots, x_{n-1}, y) =: (x, y) \end{aligned}$$

we define the sequence

$$f_n(x,y) := b(x) c_n(y) e^{(\frac{2||\rho||}{p} + is)y},$$

with an arbitrary function $b \in C_c^\infty(\varphi(N_0 \cdot x_0))$ and a (so far arbitrary) sequence $c_n \in C_c^\infty((-\infty, -(r+1)])$. Since the supports of f_n are clearly contained in the Dirichlet fundamental domain F for Γ, this sequence descends to a differentiable sequence with compact supports in M. For this reason, it suffices to perform all the calculations below in the universal covering space X of M.

We begin with the calculation of $\Delta_X f_n$ using the formula for the Laplace-Beltrami operator in horocyclic coordinates derived in Corollary 4.6:

$$\begin{aligned} \Delta_X f_n(x,y) \quad = \quad & -2\left(\sum_{i=1}^{n-1} e^{2\alpha_i(H)y} \frac{\partial^2}{\partial x_i^2} b(x)\right) c_n(y) e^{(\frac{2||\rho||}{p} + is)y} \\ & + 2||\rho|| b(x) \left(c_n'(y) + (\frac{2||\rho||}{p} + is) c_n(y)\right) e^{(\frac{2||\rho||}{p} + is)y} \\ - b(x) & \left(c_n''(y) + 2(\frac{2||\rho||}{p} + is) c_n'(y) + (\frac{2||\rho||}{p} + is)^2 c_n(y)\right) e^{(\frac{2||\rho||}{p} + is)y}. \end{aligned}$$

This leads to

$$\begin{aligned}\Delta_X f_n(x,y) &= -2\left(\sum_{i=1}^{n-1} e^{2\alpha_i(H)y}\frac{\partial^2}{\partial x_i^2}b(x)\right)c_n(y)e^{(\frac{2||\rho||}{p}+is)y}\\ &\quad + b(x)\left(zc_n(y)+(2||\rho||-2(\frac{2||\rho||}{p}+is))c_n'(y)-c_n''(y)\right)e^{(\frac{2||\rho||}{p}+is)y}.\end{aligned}$$

Therefore, we obtain

$$\begin{aligned}\Delta_X f_n(x,y)-zf_n(x,y) &= -2\left(\sum_{i=1}^{n-1} e^{2\alpha_i(H)y}\frac{\partial^2}{\partial x_i^2}b(x)\right)c_n(y)e^{(\frac{2||\rho||}{p}+is)y}\\ &\quad + b(x)\left((2||\rho||-\frac{4||\rho||}{p}-2is))c_n'(y)-c_n''(y)\right)e^{(\frac{2||\rho||}{p}+is)y}.\end{aligned}$$

Recall the volume form (cf. Corollary 4.5)

$$\left(\frac{1}{2}\right)^{(n-1)/2} e^{-2||\rho||y}\,dxdy$$

of the rank-1 symmetric space X. In the subsequent lines we estimate the L^p norm of $\Delta_X f_n - zf_n$:

$$\begin{aligned}||\Delta_X f_n - zf_n||_{L^p} &\leq C_1\sum_{i=1}^{n-1}\left(\int_X\left|e^{2\alpha_i(H)y}\frac{\partial^2}{\partial x_i^2}b(x)\right|^p|c_n(y)|^p\,dxdy\right)^{\frac{1}{p}}\\ &\quad + C_2\left(\int_X |b(x)|^p|c_n'(y)|^p\,dxdy\right)^{\frac{1}{p}}\\ &\quad + C_3\left(\int_X |b(x)|^p|c_n''(y)|^p\,dxdy\right)^{\frac{1}{p}}\\ &= C_1\sum_{i=1}^{n-1}\left(\int\left|\frac{\partial^2}{\partial x_i^2}b(x)\right|^p\,dx\right)^{\frac{1}{p}}\left(\int_{-\infty}^{\infty} e^{2p\alpha_i(H)y}|c_n(y)|^p\,dy\right)^{\frac{1}{p}}\\ &\quad + C_2\left(\int |b(x)|^p\,dx\right)^{\frac{1}{p}}\left(\int_{-\infty}^{\infty}|c_n'(y)|^p\,dy\right)^{\frac{1}{p}}\\ &\quad + C_3\left(\int |b(x)|^p\,dx\right)^{\frac{1}{p}}\left(\int_{-\infty}^{\infty}|c_n''(y)|^p\,dy\right)^{\frac{1}{p}}\end{aligned}$$

for some positive constants C_1, C_2 and C_3 (only depending on z and p). We also have

$$||f_n||_{L^p} = \left(\int |b(x)|^p\,dx\right)^{\frac{1}{p}}\left(\int_{-\infty}^{\infty}|c_n(y)|^p\,dy\right)^{\frac{1}{p}}.$$

We need to construct a sequence of functions c_n such that for increasing n these functions become flatter and flatter, and the supports tend to *negative* infinity (because of the term $e^{2p\alpha_i(H)y}$). More precisely, we choose a function $\psi \in C_c^\infty(\mathbb{R}), \psi \neq 0$, with $\operatorname{supp}(\psi) \subset (-2,-1)$ and a sequence of positive real numbers r_n such that $r_n \to \infty$ (for $n \to \infty$). We define

$$c_n(y) := \psi\left(\frac{y}{r_n}\right).$$

Since $\operatorname{supp}(c_n)$ is a subset of $(-2r_n, -r_n)$, we have $c_n \in C_c^\infty((-\infty, -(r+1)])$ for sufficiently large n. We claim that for any choice of $b \in C_c^\infty(\varphi(N_0 \cdot x_0))$ it follows

$$\frac{||\Delta_X f_n - z f_n||_{L^p}}{||f_n||_{L^p}} \to 0 \qquad (n \to \infty).$$

To see this, we compute

$$\begin{aligned}\int_{-\infty}^{\infty} e^{2p\alpha_i(H)y}|c_n(y)|^p\,dy &= r_n \int_{-2}^{-1} e^{2p\alpha_i(H)r_n u}|\psi(u)|^p\,du \\ &\leq r_n \max|\psi(u)|^p \int_{-2}^{-1} e^{2p\alpha_i(H)r_n u}\,du \\ &\leq \frac{\max|\psi(u)|^p}{2p\alpha_i(H)} e^{-2p\alpha_i(H)r_n},\end{aligned}$$

and

$$\begin{aligned}\int_{-\infty}^{\infty} |c_n(y)|^p\,dy &= r_n \int_{-2}^{-1} |\psi(u)|^p\,du, \\ \int_{-\infty}^{\infty} |c_n'(y)|^p\,dy &= r_n^{1-p} \int_{-2}^{-1} |\psi'(u)|^p\,du, \\ \int_{-\infty}^{\infty} |c_n''(y)|^p\,dy &= r_n^{1-2p} \int_{-2}^{-1} |\psi''(u)|^p\,du.\end{aligned}$$

In conclusion

$$\frac{||\Delta_X f_n - z f_n||_{L^p}}{||f_n||_{L^p}} \leq \sum_{i=1}^{n-1} c_i \frac{e^{-2\alpha_i(H)r_n}}{r_n} + \tilde{c}_1 \frac{1}{r_n} + \tilde{c}_2 \frac{1}{r_n^2},$$

with positive constants $c_i, \tilde{c}_1$ and $\tilde{c}_2$ (depending only on p, z, and the function b). Since the right hand side converges to 0 if $n \to \infty$, the proof is complete. □

Corollary 4.28. *Let $X = G/K$ denote a symmetric space of non-compact type with* $\operatorname{rank}(X) = 1$. *Assume that $\Gamma \subset G$ is a small subgroup. Then the following holds for the L^2-spectrum of the locally symmetric space $M = \Gamma\backslash X$:*

$$\sigma(\Delta_{M,2}) = [||\rho||^2, \infty).$$

Proof. The inclusion $[||\rho||^2, \infty) \subset \sigma(\Delta_{M,2})$ is a special case of the previous proposition. Since the bottom of the L^2-spectrum $\lambda_0(M)$ equals $||\rho||^2$ (cf. Theorem 3.18), the assertion follows. □

Lemma 4.29. *Let $X = G/K$ denote a symmetric space of non-compact type with $\dim X \geq 3$, $\Gamma \subset G$ a discrete, freely acting subgroup of G, and $M = \Gamma\backslash X$ the respective locally symmetric space. Assume further, that M has bounded geometry. If $1 \leq p \leq q < \infty$, we have*

$$\mathrm{e}^{-t\Delta_M}\Delta_{M,p} \subset \Delta_{M,q}\,\mathrm{e}^{-t\Delta_M}.$$

Proof. We first claim that

$$\mathrm{e}^{-t\Delta_M} : L^p(M) \to L^q(M)$$

are bounded linear operators. To see this, we notice that the general fact

$$\int k(t,\tilde{x},\tilde{y})\, dvol(\tilde{y}) \leq 1$$

implies the boundedness of $\mathrm{e}^{-t\Delta_M} : L^\infty(M) \to L^\infty(M)$ as well as the boundedness of $\mathrm{e}^{-t\Delta_M} : L^1(M) \to L^1(M)$. From Corollary 3.4 it follows that the Poincaré series $P(s;x,x)$ is bounded if $s > 2||\rho||$ since M has bounded geometry. In view of the heat kernel estimates in Corollary 3.22 we can conclude that (for fixed t) the heat kernel $k(t,\tilde{x},\tilde{y})$ is bounded. Therefore, the operators $\mathrm{e}^{-t\Delta_M} : L^1(M) \to L^\infty(M)$ are bounded. An application of the Riesz-Thorin interpolation theorem finishes the proof of our first claim.
To proof the lemma, we choose an $f \in \mathrm{dom}(\Delta_{M,p}) = \mathrm{dom}(\mathrm{e}^{-t\Delta_M}\Delta_{M,p})$. We obtain

$$||\frac{1}{s}(\mathrm{e}^{-s\Delta_M}\mathrm{e}^{-t\Delta_M}f - \mathrm{e}^{-t\Delta_M}f) - \mathrm{e}^{-t\Delta_M}\Delta_{M,p}f||_{L^q} \leq$$
$$C\,||\frac{1}{s}(\mathrm{e}^{-s\Delta_M}f - f) - \Delta_{M,p}f||_{L^p} \to 0 \quad (s \to 0^+).$$

Therefore, the function $\mathrm{e}^{-t\Delta_M}f$ is contained in the domain of $\Delta_{M,q}$ and we also have the equality

$$\mathrm{e}^{-t\Delta_M}\Delta_{M,p}f = \Delta_{M,q}\,\mathrm{e}^{-t\Delta_M}f.$$

□

Proposition 4.30. *Let $X = G/K$ denote a symmetric space of non-compact type with $\dim X \geq 3$, $\Gamma \subset G$ a discrete, freely acting subgroup of G, and $M = \Gamma\backslash X$ the respective locally symmetric space. Assume further, that M has bounded geometry. If $1 \leq p \leq q \leq 2$, we have the inclusion*

$$\sigma(\Delta_{M,q}) \subset \sigma(\Delta_{M,p}).$$

Proof. The statement of the proposition is obviously equivalent to the reverse inclusion for the respective resolvent sets:

$$\rho(\Delta_{M,p}) \subset \rho(\Delta_{M,q}).$$

Recall Lemma 4.29:

$$\mathrm{e}^{-t\Delta_M} \Delta_{M,p} \subset \Delta_{M,q}\, \mathrm{e}^{-t\Delta_M}.$$

The proof now follows as in Proposition 3.1 in [44] or Proposition 2.1 in [43]. For the sake of completeness we work out the details.
We are going to show that for $\lambda \in \rho(\Delta_{M,p}) \cap \rho(\Delta_{M,q})$ the resolvents coincide on $L^p(M) \cap L^q(M)$. From Lemma 4.29 above, we conclude for these λ

$$\begin{aligned} (\lambda - \Delta_{M,q})^{-1}\, \mathrm{e}^{-t\Delta_M} &= (\lambda - \Delta_{M,q})^{-1}\, \mathrm{e}^{-t\Delta_M} (\lambda - \Delta_{M,p})(\lambda - \Delta_{M,p})^{-1} \\ &= (\lambda - \Delta_{M,q})^{-1} (\lambda - \Delta_{M,q})\, \mathrm{e}^{-t\Delta_M} (\lambda - \Delta_{M,p})^{-1} \\ &= \mathrm{e}^{-t\Delta_M} (\lambda - \Delta_{M,p})^{-1}, \end{aligned} \tag{4.3}$$

where the equality is meant between bounded operators from $L^p(M)$ to $L^q(M)$. If $t \to 0$, we obtain

$$(\lambda - \Delta_{M,q})^{-1}|_{L^p \cap L^q} = (\lambda - \Delta_{M,p})^{-1}|_{L^p \cap L^q}.$$

For $\frac{1}{p} + \frac{1}{p'} = 1$ and $\lambda \in \rho(\Delta_{M,p}) = \rho(\Delta_{M,p'})$ we have by the preceding calculation

$$(\lambda - \Delta_{M,p'})^{-1}|_{L^p \cap L^{p'}} = (\lambda - \Delta_{M,p})^{-1}|_{L^p \cap L^{p'}}.$$

The Riesz-Thorin interpolation theorem implies that $(\lambda - \Delta_{M,p})^{-1}$ is bounded if considered as an operator R_λ on $L^q(M)$.
In the remainder of the proof we show that R_λ coincides with $(\lambda - \Delta_{M,q})^{-1}$ and hence $\rho(\Delta_{M,p}) \subset \rho(\Delta_{M,q})$. Notice, that (4.3) implies

$$(\lambda - \Delta_{M,q})\mathrm{e}^{-t\Delta_M} (\lambda - \Delta_{M,p})^{-1} f = \mathrm{e}^{-t\Delta_M} f,$$

for all $f \in L^q(M) \cap L^p(M)$. Since $\Delta_{M,q}$ is a closed operator, we obtain for $t \to 0$ the limit

$$(\lambda - \Delta_{M,q}) R_\lambda f = f.$$

As $L^q(M) \cap L^p(M)$ is dense in $L^q(M)$ and $\Delta_{M,q}$ is closed, it follows $(\lambda - \Delta_{M,q}) R_\lambda f = f$ for all $f \in L^q(M)$. Therefore, $(\lambda - \Delta_{M,q})$ is onto. If we assume that $(\lambda - \Delta_{M,q})$ is not one-to-one, λ would be an eigenvalue of $\Delta_{M,q}$. But since the semigroup $\mathrm{e}^{-t\Delta_M}$ is strongly continuous (on any $L^p(M)$) it would follow from Lemma 4.29 that there is a $t > 0$ such that $\mathrm{e}^{-t\Delta_M} f \neq 0$ is an eigenfunction with eigenvalue λ of $\Delta_{M,p'}$, and this contradicts $\lambda \in \rho(\Delta_{M,p}) = \rho(\Delta_{M,p'})$. We finally obtain $R_\lambda = (\lambda - \Delta_{M,q})^{-1}$. □

We are now prepared to prove Theorem 4.2:

Proof of Theorem 4.2. In view of Proposition 4.25 we need only show the inclusion

$$P_p \subset \sigma(\Delta_{M,p}), \qquad p \in [1,\infty).$$

But this follows for $p \in [1,2]$ from Proposition 4.26 and Proposition 4.30 by observing

$$P_p = \bigcup_{q \in [p,2]} \partial P_q.$$

The remaining cases $p \geq 2$ follow by duality as the parabolic regions P_p and $P_{p'}$ coincide if $\frac{1}{p} + \frac{1}{p'} = 1$. □

Chapter 5.

Locally Symmetric Spaces with $\mathbb{Q}$-Rank 1

5.1. Preliminaries

In Section 5.2 below we examine the L^p-spectrum of a locally symmetric space $\Gamma\backslash X = \Gamma\backslash G/K$ where Γ denotes an arithmetic, non-uniform lattice with $\mathbb{Q}$-rank 1 that acts freely on a symmetric space X of non-compact type.
In this section we provide a short introduction (without proofs) into the theory of *arithmetic groups*, the notions of *$\mathbb{Q}$-Rank* and *$\mathbb{R}$-Rank* of algebraic groups, *Siegel sets* and *reduction theory*. Our main references (for the proofs) are [5, 8, 81] and the unpublished notes [79]. A nice introduction and overview to these topics can also be found in the paper [48] and the book [9].

5.1.1. Arithmethic Groups, $\mathbb{Q}$-Rank and $\mathbb{R}$-Rank

Notation. Suppose $\mathbf{G} \subset GL(n, \mathbb{C})$ is an algebraic group and $S \subset \mathbb{C}$ is any subring with unit. We define

$$GL(n, S) := \{g \in GL(n, \mathbb{C}) : g \in S^{n\times n} \text{ and } \det g \in S^{\times}\}.$$

Furthermore, we denote by $\mathbf{G}(S) := \mathbf{G} \cap GL(n, S)$ the S-points of the algebraic group $\mathbf{G}$ and by $\mathbf{G}(S)^0$ the connected component of $\mathbf{G}(S)$ that contains the identity.

Let $G = \mathrm{Isom}^0(X)$ denote the connected component containing the identity of the isometry group of a symmetric space X of non-compact type. Then G is a non-compact semi-simple Lie group with trivial center (cf. [29], Proposition 2.1.1), and we can find a connected, semi-simple algebraic group $\mathbf{G} \subset GL(n, \mathbb{C})$ defined over

$\mathbb{Q}$ such that the groups G and $\mathbf{G}(\mathbb{R})^0$ are isomorphic as Lie groups (cf. Proposition 1.14.6 in [29]).

Let us denote by $\mathbf{T}_{\mathbb{K}} \subset \mathbf{G}$ ($\mathbb{K} = \mathbb{R}$ or $\mathbb{K} = \mathbb{Q}$) a maximal $\mathbb{K}$-split algebraic torus. Remember that we call a closed, connected subgroup $\mathbf{T}$ of $\mathbf{G}$ a *torus* if $\mathbf{T}$ is diagonalizable over $\mathbb{C}$, or equivalently if $\mathbf{T}$ is abelian and every element of $\mathbf{T}$ is semi-simple. Such a torus $\mathbf{T}$ is called $\mathbb{R}$-split if $\mathbf{T}$ is diagonalizable over $\mathbb{R}$ and $\mathbb{Q}$-split if $\mathbf{T}$ is defined over $\mathbb{Q}$ and diagonalizable over $\mathbb{Q}$.

All maximal $\mathbb{K}$-split tori in $\mathbf{G}$ are conjugate under $\mathbf{G}(\mathbb{K})$, and we call their common dimension $\mathbb{K}$-*rank* of $\mathbf{G}$. This definition can be used to define the $\mathbb{R}$-rank of the Lie group G and it turns out that the $\mathbb{R}$-rank of G coincides with the rank of the symmetric space X, i.e. the dimension of a maximal flat subspace in X. However, it is impossible to define in this way the $\mathbb{Q}$-rank of a semi-simple Lie group G since different embeddings of G in $GL(n,\mathbb{R})$ may indeed yield different values for the $\mathbb{Q}$-rank. This problem does not occur if we fix an arithmetic lattice in G. Since we are only interested in *non-uniform* lattices Γ, we may define arithmetic lattices in the following way (cf. Corollary 6.1.10 in [81] and its proof):

Definition 5.1. *A lattice $\Gamma \subset G$ in a connected semi-simple Lie group G with trivial center and no compact factors is called* arithmetic *if there are*

(i) *a semi-simple algebraic group $\mathbf{G} \subset GL(n,\mathbb{C})$ defined over $\mathbb{Q}$ and*

(ii) *an isomorphism*
$$\varphi : \mathbf{G}(\mathbb{R})^0 \to G$$

such that $\varphi(\mathbf{G}(\mathbb{Z}) \cap \mathbf{G}(\mathbb{R})^0)$ and Γ are commensurable, i.e. $\varphi(\mathbf{G}(\mathbb{Z}) \cap \mathbf{G}(\mathbb{R})^0) \cap \Gamma$ has finite index in both $\varphi(\mathbf{G}(\mathbb{Z}) \cap \mathbf{G}(\mathbb{R})^0)$ and Γ.

For the general definition of arithmetic lattices see [81], Definition 6.1.1.

A well-known and fundamental result due to G. A. Margulis ensures that this is usually the only way to obtain a lattice (for a proof see [60] or [81]):

Theorem 5.2. (Margulis' Arithmeticity Theorem). *Let G be a connected, semi-simple Lie group with trivial center, no compact factors and* $\mathbb{R}$-rank$(G) \geq 2$. *If $\Gamma \subset G$ is an irreducible lattice then Γ is arithmetic.*

Further results due to K. Corlette (cf. [17]) and M. Gromov & R. Schoen (cf. [39]) extended this result to all connected semi-simple Lie groups with trivial center except $SO(1,n)$ and $SU(1,n)$. In $SO(1,n)$ (for all $n \in \mathbb{N}$) and in $SU(1,n)$ (for $n = 2, 3$) actually non-arithmetic lattices are known to exist (cf. for example [38] and [60]).

Definition 5.3. ($\mathbb{Q}$-rank of an arithmetic lattice). *Suppose $\Gamma \subset G$ is an arithmetic lattice in a connected semi-simple Lie group G with trivial center and no compact factors. Then* $\mathbb{Q}$-rank(Γ) *is by definition the* $\mathbb{Q}$-rank *of $\mathbf{G}$, where $\mathbf{G}$ is an algebraic group as in Definition 5.1.*

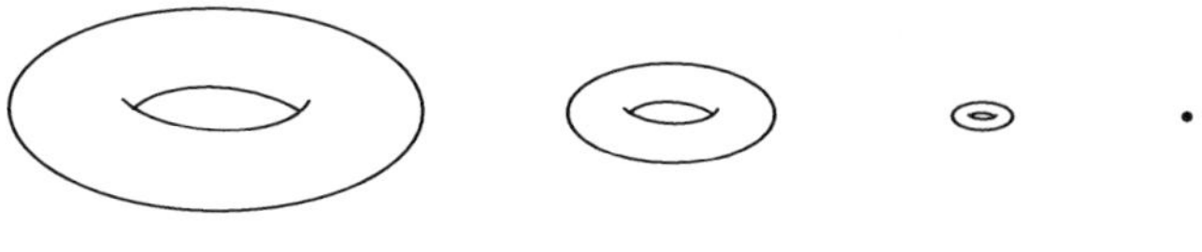

Figure 5.1.: Tangent Cone at Infinity of a Compact Space.

The theory of algebraic groups shows that the definition of the $\mathbb{Q}$-rank of an arithmetic lattice does not depend on the choice of the algebraic group $\mathbf{G}$ in Definition 5.1. A proof of this fact can be found in Chapter 9 of [79] (cf. Corollary 9.12).

We already mentioned a geometric characterization of the $\mathbb{R}$-rank: The $\mathbb{R}$-rank of a semi-simple Lie group G with trivial center coincides with the rank of the corresponding symmetric space $X = G/K$. For the $\mathbb{Q}$-rank of an arithmetic lattice Γ there is also a geometric interpretation in terms of the large scale geometry of the corresponding locally symmetric space $\Gamma\backslash G/K$:
Let us fix an arbitrary point $p \in M = \Gamma\backslash X$. The *tangent cone at infinity* of M is the (pointed) Gromov-Hausdorff limit of the sequence $(M, p, \frac{1}{n}d_M)$ of pointed metric spaces. Heuristically speaking, this means that we are looking at the locally symmetric space M from farther and farther away. The precise definition can be found in Chapter 10 of [66]. We have the following geometric characterization of $\mathbb{Q}$-rank(Γ). For a proof see [40, 58] or [79].

Theorem 5.4. *Let $X = G/K$ denote a symmetric space of non-compact type and $\Gamma \subset G$ an arithmetic lattice. The tangent cone at infinity of $\Gamma\backslash X$ is isometric to a Euclidean cone over a finite simplicial complex whose dimension is* $\mathbb{Q}$-rank(Γ).

An immediate consequence of this theorem is that $\mathbb{Q}$-rank(Γ) $= 0$ if and only if the locally symmetric space $\Gamma\backslash X$ is compact.

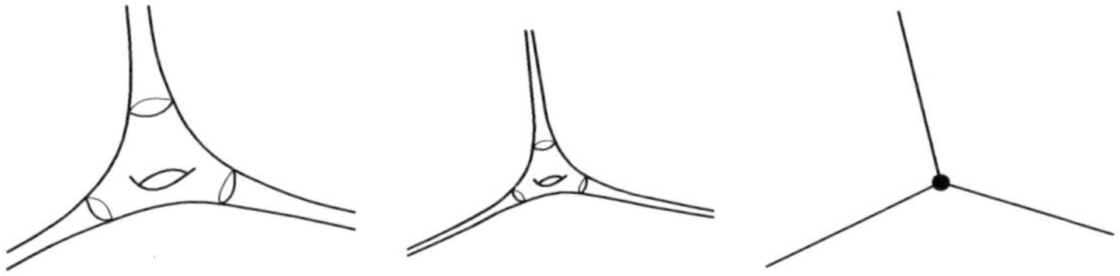

Figure 5.2.: Tangent Cone at Infinity of a $\mathbb{Q}$-Rank One Space.

5.1.2. Siegel Sets and Reduction Theory

Let us denote in this subsection by $\mathbf{G}$ again a connected, semi-simple algebraic group defined over $\mathbb{Q}$ with trivial center and by X the corresponding symmetric space of maximal compact subgroups of $G = \mathbf{G}(\mathbb{R})$.

Langlands Decomposition of Rational Parabolic Subgroups

Definition 5.5. *A closed subgroup* $\mathbf{P} \subset \mathbf{G}$ *defined over* $\mathbb{Q}$ *is called* rational parabolic subgroup *if one of the following equivalent conditions holds:*

(1) $\mathbf{G}/\mathbf{P}$ *is a projective variety.*

(2) $\mathbf{P}$ *contains a maximal, connected solvable subgroup of* $\mathbf{G}$. *(These subgroups are called* Borel subgroups *of* $\mathbf{G}$.*)*

(3) $\mathbf{G}/\mathbf{P}$ *is compact.*

For any rational parabolic subgroup $\mathbf{P}$ of $\mathbf{G}$ we denote by $\mathbf{N_P}$ the unipotent radical of $\mathbf{P}$, i.e. the largest unipotent normal subgroup of $\mathbf{P}$ and by $N_\mathbf{P} := \mathbf{N_P}(\mathbb{R})$ the real points of $\mathbf{N_P}$. The *Levi quotient* $\mathbf{L_P} := \mathbf{P}/\mathbf{N_P}$ is reductive and both $\mathbf{N_P}$ and $\mathbf{L_P}$ are defined over $\mathbb{Q}$. If we denote by $\mathbf{S_P}$ the maximal $\mathbb{Q}$-split torus in the center of $\mathbf{L_P}$ and by $A_\mathbf{P} := \mathbf{S_P}(\mathbb{R})^0$ the connected component of $\mathbf{S_P}(\mathbb{R})$ containing the identity, we obtain the decomposition of $\mathbf{L_P}(\mathbb{R})$ into $A_\mathbf{P}$ and the real points $M_\mathbf{P}$ of a reductive algebraic group $\mathbf{M_P}$ defined over $\mathbb{Q}$:

$$\mathbf{L_P}(\mathbb{R}) = A_\mathbf{P} M_\mathbf{P} \cong A_\mathbf{P} \times M_\mathbf{P}.$$

After fixing a certain basepoint $x_0 \in X$, we can lift the groups $\mathbf{L_P}, \mathbf{S_P}$ and $\mathbf{M_P}$ into $\mathbf{P}$ such that their images $\mathbf{L}_{\mathbf{P},x_0}, \mathbf{S}_{\mathbf{P},x_0}$ and $\mathbf{M}_{\mathbf{P},x_0}$ are algebraic groups defined over $\mathbb{Q}$ (this is in general not true for every choice of a basepoint x_0) and give rise to the *rational Langlands decomposition* of $P := \mathbf{P}(\mathbb{R})$:

$$P \cong N_\mathbf{P} \times A_{\mathbf{P},x_0} \times M_{\mathbf{P},x_0}.$$

More precisely, this means that the map

$$P \to N_\mathbf{P} \times A_{\mathbf{P},x_0} \times M_{\mathbf{P},x_0}, \quad g \mapsto (n(g), a(g), m(g))$$

is a real analytic diffeomorphism.

Denoting by $X_{\mathbf{P},x_0}$ the *boundary symmetric space*

$$X_{\mathbf{P},x_0} := M_{\mathbf{P},x_0}/K \cap M_{\mathbf{P},x_0}$$

we obtain, since the subgroup P acts transitively on the symmetric space $X = G/K$ (we actually have $G = PK$), the following *rational horocyclic decomposition* of X:

$$X \cong N_\mathbf{P} \times A_{\mathbf{P},x_0} \times X_{\mathbf{P},x_0}.$$

More precisely, if we denote by $\tau : M_{\mathbf{P},x_0} \to X_{\mathbf{P},x_0}$ the canonical projection, we have an analytic diffeomorphism

$$\mu : N_{\mathbf{P}} \times A_{\mathbf{P},x_0} \times X_{\mathbf{P},x_0} \to X, \ (n, a, \tau(m)) \mapsto nam \cdot x_0. \tag{5.1}$$

Note, that the boundary symmetric space $X_{\mathbf{P},x_0}$ is a Riemannian product of a symmetric space of non-compact type by a Euclidean space.

For minimal rational parabolic subgroups, i.e. Borel subgroups $\mathbf{P}$, we have

$$\dim A_{\mathbf{P},x_0} = \mathbb{Q}\text{-rank}(\mathbf{G}).$$

In the following we omit the reference to the chosen basepoint x_0 in the subscripts.

$\mathbb{Q}$-Roots

For the results in this subsection we refer also to [10] and [6].

Let us fix some *minimal* rational parabolic subgroup $\mathbf{P}$ of $\mathbf{G}$. We denote in the following by $\mathfrak{g}$, $\mathfrak{a}_{\mathbf{P}}$, and $\mathfrak{n}_{\mathbf{P}}$ the Lie algebras of the (real) Lie groups $G, A_{\mathbf{P}}$, and $N_{\mathbf{P}}$ defined above. Associated with the pair $(\mathfrak{g}, \mathfrak{a}_{\mathbf{P}})$ there is – similar to Section 1.4 – a system $\Phi(\mathfrak{g}, \mathfrak{a}_{\mathbf{P}})$ of so-called *$\mathbb{Q}$-roots*. If we define for $\alpha \in \Phi(\mathfrak{g}, \mathfrak{a}_{\mathbf{P}})$ the *root spaces*

$$\mathfrak{g}_\alpha := \{Z \in \mathfrak{g} : \mathrm{ad}(H)(Y) = \alpha(H)(Y) \text{ for all } H \in \mathfrak{a}_{\mathbf{P}}\},$$

we have the root space decomposition

$$\mathfrak{g} = \mathfrak{g}_0 \oplus \bigoplus_{\alpha \in \Phi(\mathfrak{g}, \mathfrak{a}_{\mathbf{P}})} \mathfrak{g}_\alpha,$$

where $\mathfrak{g}_0$ is the Lie algebra of $Z(\mathbf{S_P}(\mathbb{R}))$. Furthermore, the minimal rational parabolic subgroup $\mathbf{P}$ defines an ordering of $\Phi(\mathfrak{g}, \mathfrak{a}_{\mathbf{P}})$ such that

$$\mathfrak{n}_{\mathbf{P}} = \bigoplus_{\alpha \in \Phi^+(\mathfrak{g}, \mathfrak{a}_{\mathbf{P}})} \mathfrak{g}_\alpha.$$

The root spaces $\mathfrak{g}_\alpha, \mathfrak{g}_\beta$ to distinct positive roots $\alpha, \beta \in \Phi^+(\mathfrak{g}, \mathfrak{a}_{\mathbf{P}})$ are orthogonal with respect to the Killing form:

$$B(\mathfrak{g}_\alpha, \mathfrak{g}_\beta) = \{0\}.$$

In analogy to Section 1.4 we define

$$\rho_{\mathbf{P}} := \sum_{\alpha \in \Phi^+(\mathfrak{g}, \mathfrak{a}_{\mathbf{P}})} m_\alpha \alpha,$$

where m_α is the multiplicity of the positive root α, i.e. the dimension of the respective root space.

The elements of $\Phi(\mathfrak{g}, \mathfrak{a}_{\mathbf{P}})$ are differentials of characters of the maximal $\mathbb{Q}$-split torus $\mathbf{S_P}$. For convenience, we identify the $\mathbb{Q}$-roots with characters. If restricted to $A_{\mathbf{P}}$ we denote therefore the values of these characters by $\alpha(a), (a \in A_{\mathbf{P}}, \alpha \in \Phi(\mathfrak{g}, \mathfrak{a}_{\mathbf{P}}))$ which is defined by

$$\alpha(a) := \exp \alpha(\log a).$$

Siegel Sets

Since we will consider in the succeeding section only (non-uniform) arithmetic lattices Γ with $\mathbb{Q}$-rank$(\Gamma) = 1$, we restrict ourselves from now on to the case

$$\mathbb{Q}\text{-rank}(\mathbf{G}) = 1.$$

For these groups we summarize several facts in the next lemma.

Lemma 5.6. *Assume* $\mathbb{Q}$-rank$(\mathbf{G}) = 1$. *Then the following holds:*

(1) *For any proper rational parabolic subgroup* $\mathbf{P}$ *of* $\mathbf{G}$, *we have* $\dim A_{\mathbf{P}} = 1$.

(2) *All proper rational parabolic subgroups are minimal.*

(3) *The set* $\Phi^{++}(\mathfrak{g}, \mathfrak{a}_{\mathbf{P}})$ *of simple positive* $\mathbb{Q}$*-roots contains only a single element:*

$$\Phi^{++}(\mathfrak{g}, \mathfrak{a}_{\mathbf{P}}) = \{\alpha\}.$$

For any rational parabolic subgroup $\mathbf{P}$ of $\mathbf{G}$ and any $t > 1$, we define

$$A_{\mathbf{P},t} := \{a \in A_{\mathbf{P}} : \alpha(a) > t\},$$

where α denotes the unique root in $\Phi^{++}(\mathfrak{g}, \mathfrak{a}_{\mathbf{P}})$.
If we choose $a_0 \in A_{\mathbf{P}}$ with the property $\alpha(a_0) = t$, the set $A_{\mathbf{P},t}$ is just a shift of the positive Weyl chamber $A_{\mathbf{P},1}$ by a_0:

$$A_{\mathbf{P},t} = A_{\mathbf{P},1} a_0.$$

Before we define *Siegel sets*, we recall the horocyclic decomposition of the symmetric space $X = G/K$:

$$X \cong N_{\mathbf{P}} \times A_{\mathbf{P}} \times X_{\mathbf{P}}.$$

Definition 5.7. *Let* $\mathbf{P}$ *denote a rational parabolic subgroup of the algebraic group* $\mathbf{G}$ *with* $\mathbb{Q}$-rank$(\mathbf{G}) = 1$. *For any bounded set* $\omega \subset N_{\mathbf{P}} \times X_{\mathbf{P}}$ *and any* $t > 1$, *the set*

$$\mathcal{S}_{\mathbf{P},\omega,t} := \omega \times A_{\mathbf{P},t} \subset X$$

is called Siegel set.

Precise Reduction Theory

We fix an arithmetic lattice $\Gamma \subset G = \mathbf{G}(\mathbb{R})$ in the algebraic group $\mathbf{G}$ with $\mathbb{Q}$-rank$(\mathbf{G}) = 1$. Recall, that by a well known result due to A. Borel and Harish-Chandra there are only finitely many Γ-conjugacy classes of minimal parabolic subgroups (see e.g. [5]). Using the Siegel sets defined above, we can state the *precise reduction theory* in the $\mathbb{Q}$-rank one case as follows:

Theorem 5.8. *Let* $\mathbf{G}$ *denote a semi-simple algebraic group defined over* $\mathbb{Q}$ *with* $\mathbb{Q}$-rank$(\mathbf{G}) = 1$ *and* Γ *an arithmetic lattice in* G. *We further denote by* $\mathbf{P}_1, \ldots, \mathbf{P}_k$ *representatives of the* Γ*-conjugacy classes of all rational proper (i.e. minimal) parabolic subgroups of* $\mathbf{G}$. *Then there exist a bounded set* $\Omega_0 \subset X$ *and Siegel sets* $\omega_j \times A_{\mathbf{P}_j,t_j}$ $(j = 1, \ldots, k)$ *such that the following holds:*

(1) *Under the canonical projection* $\pi : X \to \Gamma\backslash X$ *each Siegel set* $\omega_j \times A_{\mathbf{P}_j,t_j}$ *is mapped injectively into* $\Gamma\backslash X$ $(i = 1, \ldots, k)$.

(2) *The image of* ω_j *in* $(\Gamma \cap P_j)\backslash N_{\mathbf{P}_j} \times X_{\mathbf{P}_j}$ *is compact* $(j = 1, \ldots, k)$.

(3) *We have the following disjoint decomposition of* $\Gamma\backslash X$:

$$\Gamma\backslash X = \pi(\Omega_0) \cup \coprod_{j=1}^{k} \pi(\omega_j \times A_{\mathbf{P}_j,t_j}).$$

In particular, the subset $\Omega_0 \cup \coprod_{j=1}^{k} \omega_j \times A_{\mathbf{P}_j,t_j}$ *is an exact fundamental domain for* Γ.

Geometrically this means that each set $\pi(\omega_j \times A_{\mathbf{P}_j,t_j})$ corresponds to one cusp of the locally symmetric space $\Gamma\backslash X$ and the numbers t_j are chosen large enough such that these sets do not overlap. Then the bounded set $\pi(\Omega_0)$ is just the complement of $\coprod_{j=1}^{k} \pi(\omega_j \times A_{\mathbf{P}_j,t_j})$.

Figure 5.3.: Disjoint Decomposition of a $\mathbb{Q}$-rank-1 Space.

Since in the case $\mathbb{Q}$-rank$(\mathbf{G}) = 1$ all rational proper parabolic subgroups are minimal, these subgroups are conjugate under $\mathbf{G}(\mathbb{Q})$ (cf. [5], Theorem 11.4). Therefore, the root systems $\Phi(\mathfrak{g}, \mathfrak{a}_{\mathbf{P}_j})$ with respect to the rational proper parabolic subgroups $\mathbf{P}_j$, $j = 1 \ldots k$, are canonically isomorphic (cf. [5], 11.9) and moreover, we can conclude $||\rho_{\mathbf{P}_1}|| = \ldots = ||\rho_{\mathbf{P}_k}||$.

5.1.3. Rational Horocyclic Coordinates

For all $\alpha \in \Phi^+(\mathfrak{g}, \mathfrak{a}_{\mathbf{P}})$ we define on $\mathfrak{n}_{\mathbf{P}} = \bigoplus_{\alpha \in \Phi^+(\mathfrak{g}, \mathfrak{a}_{\mathbf{P}})} \mathfrak{g}_\alpha$ a left invariant bilinear form h_α by

$$h_\alpha := \begin{cases} \langle \cdot, \cdot \rangle, & \text{on } \mathfrak{g}_\alpha \\ 0, & \text{else,} \end{cases}$$

where $\langle\cdot,\cdot\rangle$ denotes the scalar product defined by means of the Killing form. We then have (cf. Proposition 1.6 in [6] and Proposition 4.3 in [7]):

Proposition 5.9. (a) *For any* $x = (n, \tau(m), a) \in X \cong N_{\mathbf{P}} \times X_{\mathbf{P}} \times A_{\mathbf{P}}$ *the tangent spaces at* x *to the submanifolds* $\{n\} \times X_{\mathbf{P}} \times \{a\}$, $\{n\} \times \{\tau(m)\} \times A_{\mathbf{P}}$, *and* $N_{\mathbf{P}} \times \{\tau(m)\} \times \{a\}$ *are mutually orthogonal.*

(b) *The pullback* $\mu^* g$ *of the metric* g *on* X *to* $N_{\mathbf{P}} \times X_{\mathbf{P}} \times A_{\mathbf{P}}$ *is given by*

$$ds^2_{(n,a,\tau(m))} = \frac{1}{2} \sum_{\alpha \in \Phi^+(\mathfrak{g},\mathfrak{a}_{\mathbf{P}})} \mathrm{e}^{-2\alpha(\log a)} h_\alpha \oplus d(\tau(m))^2 \oplus da^2.$$

If we choose (as in the preceding chapter) orthonormal bases $\{N_1, \ldots, N_r\}$ of $\mathfrak{n}_{\mathbf{P}}$, $\{Y_1, \ldots, Y_l\}$ of some tangent space $T_{\tau(m)} X_{\mathbf{P}}$ and $H \in \mathfrak{a}_{\mathbf{P}}^+$ with $||H|| = 1$, we obtain *rational horocyclic coordinates*

$$\begin{aligned} \varphi \; : \; & N_{\mathbf{P}} \times X_{\mathbf{P}} \times A_{\mathbf{P}} \to \mathbb{R}^r \times \mathbb{R}^l \times \mathbb{R}, \\ & \left(\exp(\sum_{j=1}^{r} x_j N_j), \exp(\sum_{j=1}^{l} x_{j+r} Y_j), \exp(yH) \right) \mapsto (x_1, \ldots, x_{r+l}, y). \end{aligned}$$

In the following, we will abbreviate $(x_1, \ldots, x_{r+l}, y)$ as (x, y). The representation of the metric ds^2 with respect to these coordinates is given by the matrix

$$(g_{ij})_{i,j}(n, \tau(m), a) = \begin{pmatrix} \begin{matrix} \frac{1}{2}\mathrm{e}^{-2\alpha_1(\log a)} & & \\ & \ddots & \\ & & \frac{1}{2}\mathrm{e}^{-2\alpha_r(\log a)} \end{matrix} & 0 \\ 0 & \begin{matrix} h_{km} & \\ & 1 \end{matrix} \end{pmatrix}$$

where the positive roots $\alpha_i \in \Phi^+(\mathfrak{g}, \mathfrak{a}_{\mathbf{P}})$ appear according to their multiplicity and the $(l \times l)$-submatrix $(h_{km})_{k,m=1}^{l}$ represents the metric $d(\tau(m))^2$ on the boundary symmetric space $X_{\mathbf{P}}$.

Corollary 5.10. *The volume form of* $N_{\mathbf{P}} \times X_{\mathbf{P}} \times A_{\mathbf{P}}$ *with respect to rational horocyclic coordinates is given by*

$$\begin{aligned} \sqrt{\det(g_{ij})(n, \tau(m), a)}\, dxdy &= \left(\frac{1}{2}\right)^{r/2} \sqrt{\det(h_{km}(\tau(m))}\; \mathrm{e}^{-2\rho_{\mathbf{P}}(\log a)} dxdy \\ &= \left(\frac{1}{2}\right)^{r/2} \sqrt{\det(h_{km}(\tau(m))}\; \mathrm{e}^{-2y\rho_{\mathbf{P}}(H)} dxdy \\ &= \left(\frac{1}{2}\right)^{r/2} \sqrt{\det(h_{km}(\tau(m))}\; \mathrm{e}^{-2y||\rho_{\mathbf{P}}||} dxdy, \end{aligned}$$

where $\log a = yH$.

A straightforward calculation yields:

Corollary 5.11. *The Laplacian on $N_{\mathbf{P}} \times X_{\mathbf{P}} \times A_{\mathbf{P}}$ in rational horocyclic coordinates is*

$$\Delta \;=\; -2\sum_{j=1}^{r} e^{2\alpha_j}\frac{\partial^2}{\partial x_j^2} - \sum_{j=r+1}^{r+l}\frac{\partial^2}{\partial x_j^2} - \frac{\partial^2}{\partial y^2} + 2||\rho_{\mathbf{P}}||\frac{\partial}{\partial y} + \Delta_{X_{\mathbf{P}}}, \qquad (5.2)$$

where $\Delta_{X_{\mathbf{P}}}$ denotes the Laplacian on the boundary symmetric space $X_{\mathbf{P}}$.

5.2. L^p-Spectrum

In this section $X = G/K$ denotes a symmetric space of non-compact type whose metric coincides on $T_{eK}(G/K) \cong \mathfrak{p}$ with the Killing form on the Lie algebra $\mathfrak{g}$ of G. Furthermore, $\Gamma \subset G$ is an arithmetic, non-uniform lattice with $\mathbb{Q}$-rank$(\Gamma) = 1$.

The corresponding locally symmetric space $M = \Gamma\backslash X$ has finitely many cusps and each cusp corresponds to a Γ-conjugacy class of a minimal rational parabolic subgroup $\mathbf{P} \subset \mathbf{G}$. Let $\mathbf{P}_1, \ldots, \mathbf{P}_k$ denote representatives of the Γ-conjugacy classes. Since these subgroups are conjugate under $\mathbf{G}(\mathbb{Q})$ and the respective root systems are isomorphic (cf. Section 5.1.2), we consider in the following only the rational parabolic subgroup $\mathbf{P} := \mathbf{P}_1$. We denote by $\rho_{\mathbf{P}}$ as in the preceding section half the sum of the positive roots (counted according to their multiplicity) with respect to the pair $(P, A_{\mathbf{P}})$.

In analogy with Definition 4.24, we define for any $p \in [1, \infty)$ the parabolic region

$$P_p := \left\{ z = x + iy \in \mathbb{C} : x \geq \frac{4||\rho_{\mathbf{P}}||^2}{p}\left(1 - \frac{1}{p}\right) + \frac{y^2}{4||\rho_{\mathbf{P}}||^2(1-\frac{2}{p})^2} \right\}$$

if $p \neq 2$ and $P_2 := [||\rho_{\mathbf{P}}||^2, \infty)$.

The boundary ∂P_p of P_p is parametrized by the curve

$$\begin{aligned} \mathbb{R} \to \mathbb{C}, \quad s \mapsto \;& \frac{4||\rho_{\mathbf{P}}||^2}{p}\left(1 - \frac{1}{p}\right) + s^2 + 2i||\rho_{\mathbf{P}}||s\left(1 - \frac{2}{p}\right) \\ &= \left(\frac{2||\rho_{\mathbf{P}}||}{p} + is\right)\left(2||\rho_{\mathbf{P}}|| - \frac{2||\rho_{\mathbf{P}}||}{p} - is\right). \end{aligned}$$

Our main result in this chapter reads as follows:

Theorem 5.12. *Let $X = G/K$ denote a symmetric space of non-compact type and $\Gamma \subset G$ an arithmetic lattice with $\mathbb{Q}$-rank$(\Gamma) = 1$. If we denote by $M := \Gamma\backslash X$ the corresponding locally symmetric space, the parabolic region P_p is contained in the spectrum of $\Delta_{M,p}$, $p \in (1, \infty)$:*

$$P_p \subset \sigma(\Delta_{M,p}).$$

The picture is quite similar to the one of Chapter 4 (see e.g. Figure 4.1).

Lemma 5.13. *Let M denote a Riemannian manifold with finite volume. For $2 \le p \le q < \infty$, we have the inclusion*

$$\sigma(\Delta_{M,p}) \subset \sigma(\Delta_{M,q}).$$

Proof. Since the volume of M is finite, it follows for $1 \le p \le q \le \infty$ (by Hölder's inequality)

$$L^q(M) \hookrightarrow L^p(M),$$

i.e. $L^q(M)$ is continuously embedded in $L^p(M)$. Therefore, we obtain the boundedness of the operators

$$\mathrm{e}^{-t\Delta_M} : L^q(M) \to L^p(M), \quad p \le q.$$

It now follows as in Lemma 4.29

$$\mathrm{e}^{-t\Delta_M}\Delta_{M,q} \subset \Delta_{M,p}\,\mathrm{e}^{-t\Delta_M} \qquad (1 \le p \le q \le \infty).$$

With this result at hand we can argue as in the proof of Proposition 4.30 to prove the claim. □

Proposition 5.14. *For $1 \le p < \infty$ the boundary ∂P_p of the parabolic region P_p is contained in the approximate point spectrum of $\Delta_{M,p}$:*

$$\partial P_p \subset \sigma_{app}(\Delta_{M,p}).$$

Proof. Choose some

$$z = z(s) = \left(\frac{2||\rho_{\mathbf{P}}||}{p} + is\right)\left(2||\rho_{\mathbf{P}}|| - \frac{2||\rho_{\mathbf{P}}||}{p} - is\right) \in \partial P_p.$$

The subsequent proof uses the same ideas as the proof of Proposition 4.26, i.e. we construct a sequence f_n of differentiable functions in $L^p(X)$ with support in a fundamental domain for Γ such that

$$\frac{||\Delta_X f_n - z f_n||_{L^p}}{||f_n||_{L^p}} \to 0 \qquad (n \to \infty).$$

A fundamental domain for Γ is given by a subset of the form

$$\Omega_0 \cup \coprod_{i=1}^{k} \omega_i \times A_{\mathbf{P}_i,t_i} \subset X$$

(cf. Theorem 5.8), and each Siegel set $\omega_i \times A_{\mathbf{P}_i,t_i}$ is mapped injectively into $\Gamma\backslash X$. Furthermore, the closure of $\pi(\omega_i \times A_{\mathbf{P}_i,t_i})$ fully covers an end of $\Gamma\backslash X$ (for any $i \in \{1, \ldots, k\}$).

We take the Siegel set $\omega \times A_{\mathbf{P},t} := \omega_1 \times A_{\mathbf{P}_1,t_1}$ where $A_{\mathbf{P},t} = \{a \in A_{\mathbf{P}} : \alpha(a) > t\}$, and define a sequence f_n of smooth functions with support in $\omega \times A_{\mathbf{P},t}$ with respect to rational horocyclic coordinates by

$$f_n(x,y) := c_n(y)\mathrm{e}^{(\frac{2}{p}||\rho_{\mathbf{P}}||+is)y},$$

where $c_n \in C_c^\infty\left((\frac{\log t}{||\alpha||},\infty)\right)$. Since ω is bounded, each f_n is clearly contained in $L^p(X)$. Furthermore, the condition $\mathrm{supp}(c_n) \subset (\frac{\log t}{||\alpha||},\infty)$ ensures that the supports of the sequence f_n are contained in the Siegel set $\omega \times A_{\mathbf{P},t}$.

Using formula (5.2) for the Laplacian in rational horocyclic coordinates, we obtain after a straightforward calculation

$$\begin{aligned}\Delta_X f_n(x,y) - zf_n(x,y) &= \\ &= \left(-c_n''(y) + \left(2||\rho_{\mathbf{P}}|| - 2(\frac{2}{p}||\rho_{\mathbf{P}}|| + is)\right)c_n'(y)\right)\mathrm{e}^{(\frac{2||\rho_{\mathbf{P}}||}{p}+is)y},\end{aligned}$$

and therefore

$$\begin{aligned}||\Delta_X f_n - zf_n||_{L^p}^p &= \int_{\omega\times A_{\mathbf{P},t}} |\Delta_X f_n - zf_n|^p dvol_X \\ &= \left(\frac{1}{2}\right)^{r/2}\int_{\omega\times A_{\mathbf{P},t}} |\Delta_X f_n(x,y) - zf_n(x,y)|^p \sqrt{\det(h_{km}(\tau(m))}\,\mathrm{e}^{-2||\rho_{\mathbf{P}}||y}\,dxdy \\ &= C\int_0^\infty \left|-c_n''(y) + \left(2||\rho_{\mathbf{P}}|| - 2(\frac{2||\rho_{\mathbf{P}}||}{p} + is)\right)c_n'(y)\right|^p dy,\end{aligned}$$

where $C := \left(\frac{1}{2}\right)^{r/2}\int_\omega \sqrt{\det(h_{km}(\tau(m))}\,dx < \infty$ because $\omega \subset N_{\mathbf{P}} \times X_{\mathbf{P}}$ is bounded. This yields after an application of the triangle inequality

$$||\Delta_X f_n - zf_n||_{L^p} \leq C_1\left(\int_0^\infty |c_n''(y)|^p\,dy\right)^{1/p} + C_2\left(\int_0^\infty |c_n'(y)|^p\,dy\right)^{1/p}.$$

By an analogous calculation we obtain

$$||f_n||_{L^p} = C_3\left(\int_0^\infty |c_n(y)|^p\,dy\right)^{1/p}.$$

We choose a function $\psi \in C_c^\infty(\mathbb{R})$, not identically zero, with $\mathrm{supp}(\psi) \subset (1,2)$, a sequence $r_n > 0$ with $r_n \to \infty$ (if $n \to \infty$), and we eventually define

$$c_n(y) := \psi\left(\frac{y}{r_n}\right).$$

For large enough n, we have $\operatorname{supp}(c_n) \subset (\frac{\log t}{||\alpha||}, \infty)$. As in the preceding chapter, we obtain

$$\begin{aligned}\int_0^\infty |c_n(y)|^p\,dy &= r_n \int_1^2 |\psi(u)|^p\,du,\\ \int_0^\infty |c_n'(y)|^p\,dy_1 &= r_n^{1-p} \int_1^2 |\psi'(u)|^p\,du,\\ \int_0^\infty |c_n''(y)|^p\,dy_1 &= r_n^{1-2p} \int_1^2 |\psi''(u)|^p\,du.\end{aligned}$$

In the end, this leads to the inequality

$$\frac{||\Delta_X f_n - z f_n||_p}{||f_n||_p} \le \frac{C_4}{r_n} + \frac{C_5}{r_n^2} \longrightarrow 0 \qquad (n \to \infty),$$

where $C_4, C_5 > 0$ denote positive constants, and the proof is complete. □

Proof of Theorem 5.12. The inclusion

$$P_p \subset \sigma(\Delta_{M,p})$$

for $p \in [2,\infty)$ follows as in the proof of Theorem 4.2. But now we use Lemma 5.13, Proposition 5.14, and note

$$P_p = \bigcup_{q\in[2,p]} \partial P_q.$$

The inclusion for all $p \in (1,\infty)$ follows again by duality. □

Up to now, we considered non-uniform arithmetic lattices $\Gamma \subset G$ with $\mathbb{Q}$-rank one. We made no assumption concerning the rank of the respective symmetric space $X = G/K$ of non-compact type. However, if $\operatorname{rank}(X) = 1$, we are able to sharpen the result of Theorem 5.12 considerably. In the case $\mathbb{Q}\text{-rank}(\Gamma) = \operatorname{rank}(X) = 1$, the one dimensional abelian subgroup $A_{\mathbf{P}}$ of G (with respect to some rational minimal parabolic subgroup) defines a maximal flat subspace, i.e. a geodesic, $A_{\mathbf{P}} \cdot x_0$ of X. Hence, the $\mathbb{Q}$-roots coincide with the roots defined in Chapter 1 and for any rational minimal parabolic subgroup $\mathbf{P}$ we have in particular

$$||\rho_{\mathbf{P}}|| = ||\rho||.$$

This proves one part of the following corollary.

Corollary 5.15. *Let $X = G/K$ denote a symmetric space of non-compact type with* $\operatorname{rank}(X) = 1$. *Furthermore, $\Gamma \subset G$ denotes a non-uniform arithmetic lattice, and $M = \Gamma\backslash X$ is the corresponding locally symmetric space. Then, we have for all $p \in (1,\infty)$ the equality*

$$\sigma(\Delta_{M,p}) = \{\lambda_0, \dots, \lambda_m\} \cup P_p,$$

where $0 = \lambda_0, \dots, \lambda_m \in \left[0, ||\rho||^2\right)$ are eigenvalues of $\Delta_{M,2}$.

Proof. Langlands' theory of Eisenstein series implies (see e.g. [55] or the surveys in [48] or [9])

$$\sigma(\Delta_{M,2}) = \{\lambda_0, \dots, \lambda_m\} \cup [||\rho||^2, \infty).$$

Thus, we can apply Proposition 3.3 from [75] and obtain

$$\sigma(\Delta_{M,p}) \subset \{\lambda_0, \dots, \lambda_m\} \cup P_p.$$

As in the proof of Lemma 6 in [24] one sees that the discrete part of the L^2-spectrum $\{\lambda_0, \dots, \lambda_m\}$ is also contained in $\sigma(\Delta_{M,p})$ for any $p \in (1, \infty)$. Together with Theorem 5.12 and the remark above this concludes the proof. □

Recall, that the critical exponent $\delta(\Gamma)$ attains for lattices Γ the maximal possible value: $\delta(\Gamma) = 2||\rho||$. In Theorem 4.2 we proved a similar result in the rank one case if the critical exponent is small, i.e. $\delta(\Gamma) \leq ||\rho||$. In the case $\delta(\Gamma) \in (||\rho||, 2||\rho||)$ it is still not clear what the L^p-spectrum is but it is quite natural to conjecture that an analogous result to Corollary 5.15 holds.

Chapter 6.

Manifolds with Cusps of Rank 1

In this chapter we consider a class of Riemannian manifolds that is larger than the class of $\mathbb{Q}$-rank one locally symmetric spaces but contains all of them. This larger class consists of those manifolds which are isometric – after the removal of a compact set – to a disjoint union of rank one cusps. Manifolds with cusps of rank one were probably first introduced and studied by W. Müller (see e.g. [61]).

6.1. Definition

Recall, that we denoted by $\omega \times A_{\mathbf{P},t} \subset X$ Siegel sets of a symmetric space $X = G/K$ of non-compact type. The projection $\pi(\omega \times A_{\mathbf{P},t})$ of certain Siegel sets to a corresponding $\mathbb{Q}$-rank one locally symmetric space $\Gamma\backslash X$ is a cusp and every cusp of $\Gamma\backslash X$ is of this form (cf. Section 5.1.2).

Definition 6.1. *A Riemannian manifold is called* cusp of rank one *if it is isometric to a cusp $\pi(\omega \times A_{\mathbf{P},t})$ of a $\mathbb{Q}$*-rank *one locally symmetric space.*

Definition 6.2. *A complete Riemannian manifold M is called* manifold with cusps of rank one *if it has a decomposition*

$$M = M_0 \cup \bigcup_{j=1}^{k} M_j$$

such that the following holds:

(i) M_0 is a compact manifold with boundary.

(ii) The subsets M_j, $j \in \{0, \ldots, k\}$, are pairwise disjoint.

(iii) For each $j \in \{1, \ldots, k\}$ there exists a cusp of rank one isometric to M_j.

Such manifolds certainly have finite volume as there is only a finite number of cusps possible and every cusp of rank one has finite volume.

From Theorem 5.8 it follows that any $\mathbb{Q}$-rank one locally symmetric space is a manifold with cusps of rank one. But since we can perturb the metric on the compact manifold M_0 without leaving the class of manifolds with cusps of rank one, not every such manifold is locally symmetric. Of course, they are locally symmetric on each cusp and we can say that they are locally symmetric near infinity.

6.2. L^p-Spectrum and Geometry

Precisely as in Proposition 5.14 one sees that we can find for every cusp M_j, $j \in \{1,\ldots,k\}$ of a manifold $M = M_0 \cup \bigcup_{j=1}^k M_j$ with cusps of rank one a parabolic region $P_p^{(j)}$ such that the boundary $\partial P_p^{(j)}$ is contained in the approximate point spectrum of $\Delta_{M,p}$. Here, the parabolic regions are defined as the parabolic regions in the two preceding chapters (cf. for example Definition 4.24), where the constant $||\rho||$ or rather $||\rho_{\mathbf{P}}||$ is replaced by an analogous quantity, say $||\rho_{\mathbf{P}_j}||$, coming from the respective cusp M_j. That is to say, we have the following lemma:

Lemma 6.3. *Let M denote a manifold with cusps of rank one. Then we have for $p \in [1,\infty)$:*

$$\partial P_p^{(j)} \subset \sigma_{app}(\Delta_{M,p}).$$

Since the volume of a manifold with cusps of rank one is finite, we can apply Lemma 5.13 in order to prove (cf. the proof of Theorem 5.12) the following Theorem:

Theorem 6.4. *Let $M = M_0 \cup \bigcup_{j=1}^k M_j$ denote a manifold with cusps of rank one. Then, for $p \in (1,\infty)$, every cusp M_j defines a parabolic region $P_p^{(j)}$ that is contained in the L^p-spectrum:*

$$\bigcup_{j=1}^{k} P_p^{(j)} \subset \sigma(\Delta_{M,p}).$$

Of course, the compact submanifold M_0 contributes some discrete set to the L^p-spectrum, and 0 is always an eigenvalue as the volume of M is finite. It seems to be very likely that besides some discrete spectrum the union of the parabolic regions in Theorem 6.4 is already the complete spectrum. But at present, I do not know how to prove this result. The methods used in [24] or [75] to prove a similar result need either that the manifold is homogeneous or that the injectivity radius is bounded from below, and it is not clear how one could adapt the methods therein to our case.

Nevertheless, given the L^p-spectrum for some $p \neq 2$, we have the following geometric consequences:

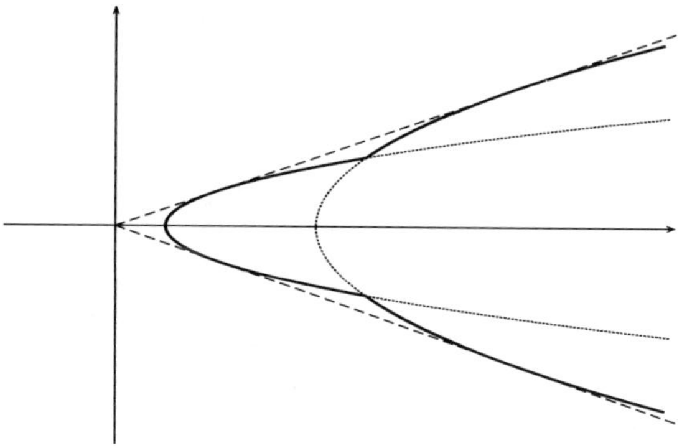

Figure 6.1.: The union of two parabolic regions $P_p^{(1)}$ and $P_p^{(2)}$ if $p \neq 2$.

Corollary 6.5. *Let $M = M_0 \cup \bigcup_{j=1}^{k} M_j$ denote a manifold with cusps of rank one such that*

$$\sigma(\Delta_{M,p}) = \{\lambda_0, \ldots, \lambda_r\} \cup P_p,$$

for some $p \neq 2$ and some parabolic region P_p. Then every cusp M_j is of the form $\pi(\omega_j \times A_{\mathbf{P}_j,t_j})$ with volume form

$$\left(\frac{1}{2}\right)^{r_j/2} \mathrm{e}^{-2yc}\, dxdy,$$

where c is a positive constant.

Proof. Since all parabolic regions $P_p^{(j)}$ induced by the cusps M_j coincide, the quantities $||\rho_{\mathbf{P}_j}||$ coincide. Therefore, we can take $c := ||\rho_{\mathbf{P}_1}||$. □

This result generalizes to the case where the continuous spectrum consists of a finite number of parabolic regions in an obvious manner.

Appendix A.

Tensor Products

We adopt the terminology used in the book [25] which serves also as our main reference in this chapter. The field $\mathbb{K}$ denotes in this part always the real numbers $\mathbb{R}$ or the complex numbers $\mathbb{C}$.

Let us first recall the definition of a tensor product of vector spaces and some basic properties.

Definition A.1. *Let E and F denote $\mathbb{K}$-vector spaces. A pair (H, Ψ_0) consisting of a $\mathbb{K}$-vector space H and a bilinear map $\Psi_0 : E \times F \to H$ is called* tensor product *of the pair (E, F) if for each $\mathbb{K}$-vector space G and each bilinear map $\Phi : E \times F \to G$ there exists a unique linear map $T : H \to G$ with $\Phi = T \circ \Psi_0$:*

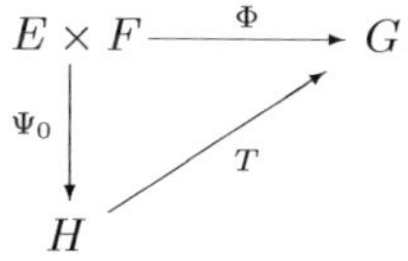

One can prove that up to isomorphisms there is a unique tensor product of the vector spaces E and F: If $(H_1, \Psi_0^{(1)})$ and $(H_2, \Psi_0^{(2)})$ are two tensor products, we can find an isomorphism $S : H_1 \to H_2$ such that $S \circ \Psi_0^{(1)} = \Psi_0^{(2)}$. We write $E \otimes F$ for the (unique) tensor product and $x \otimes y$ for the image $\Psi_0(x, y)$. Note, that we have

$$E \otimes F = \operatorname{span}\{x \otimes y : x \in E, y \in F\}.$$

If E and F are normed spaces, there are in general many possibilities to define a norm on the tensor product $E \otimes F$. Given a norm α on $E \otimes F$, we denote by $E \tilde{\otimes}_\alpha F$ the completion of the normed space $(E \otimes F, \alpha)$.

In the proof of Theorem 2.11 we claimed

$$L^p(M_1 \times M_2) \cong L^p(M_1)\tilde{\otimes}_{\triangle_p} L^p(M_2) \qquad (1 < p < \infty),$$

where $\cong$ means isometrically isomorphic. The norm $\triangle_p$ on the tensor product $L^p(M_1) \otimes L^p(M_2)$ is defined as follows: Given $\sum_{i=1}^n f_i \otimes g_i \in L^p(M_1) \otimes L^p(M_2)$, we put

$$\triangle_p\left(\sum_{i=1}^n f_i \otimes g_i\right) := \left(\int_{M_1}\int_{M_2}\left|\sum_{i=1}^n f_i(x_1)g_i(x_2)\right|^p dx_2\, dx_1\right)^{1/p}.$$

Using Fubini's theorem, we can conclude that the inclusion

$$\begin{aligned} \iota: \quad & L^p(M_1) \otimes_{\triangle_p} L^p(M_2) \to L^p(M_1 \times M_2), \\ & \textstyle\sum_{i=1}^n f_i \otimes g_i \mapsto \sum_{i=1}^n f_i g_i \end{aligned}$$

is isometric. Since the image of ι is dense in $L^p(M_1 \times M_2)$ (it containes the tensor product of the step functions in $L^p(M_1)$ and $L^p(M_2)$), the claim follows. For a more general treatment cf. Section 7 in [25].

To prove Theorem 2.11, we want to apply Theorem 5 in [68] where an appropriate spectral mapping theorem for operators of the form $A_1 \otimes I + I \otimes A_2$ on the tensor product $E\tilde{\otimes}_\alpha F$ of two Banach spaces E and F is proven. The norm α can be an arbitrary *uniform cross norm* (or *tensor norm*). For a definition cf. 12.1 in [25]. Unfortunately, $\triangle_p$ is not a uniform cross norm (cf. also 12.1 in [25]). But we are lucky, because there is a uniform cross norm g_p, which coincides with $\triangle_p$ on the tensor product $L^p(M_1) \otimes L^p(M_2)$. This is the content of Corollary 2 in 15.10 in [25]. For the definition of g_p see 12.7 therein.

Bibliography

[1] Jean-Philippe Anker & Lizhen Ji, *Heat kernel and Green function estimates on noncompact symmetric spaces*, Geom. Funct. Anal. **9** (1999), no. 6, 1035–1091.

[2] Jean-Philippe Anker & Patrick Ostellari, *The heat kernel on noncompact symmetric spaces*, Lie groups and symmetric spaces, Amer. Math. Soc. Transl. Ser. 2, vol. 210, Amer. Math. Soc., Providence, RI, 2003, pp. 27–46.

[3] Wolfgang Arendt, *Gaussian estimates and interpolation of the spectrum in L^p*, Differential Integral Equations **7** (1994), no. 5-6, 1153–1168.

[4] Robert Azencott, *Behavior of diffusion semi-groups at infinity*, Bull. Soc. Math. France **102** (1974), 193–240.

[5] Armand Borel, *Introduction aux groupes arithmétiques*, Publications de l'Institut de Mathématique de l'Université de Strasbourg, XV. Actualités Scientifiques et Industrielles, No. 1341, Hermann, Paris, 1969.

[6] Armand Borel, *Some metric properties of arithmetic quotients of symmetric spaces and an extension theorem*, J. Differential Geometry **6** (1972), 543–560.

[7] Armand Borel, *Stable real cohomology of arithmetic groups*, Ann. Sci. École Norm. Sup. (4) **7** (1974), 235–272.

[8] Armand Borel, *Linear algebraic groups*, second ed., Graduate Texts in Mathematics, vol. 126, Springer-Verlag, New York, 1991.

[9] Armand Borel & Lizhen Ji, *Compactifications of symmetric and locally symmetric spaces*, Mathematics: Theory & Applications, Birkhäuser Boston Inc., Boston, MA, 2006.

[10] Armand Borel & Jacques Tits, *Groupes réductifs*, Inst. Hautes Études Sci. Publ. Math. (1965), no. 27, 55–150.

[11] Brian H. Bowditch, *Geometrical finiteness with variable negative curvature*, Duke Math. J. **77** (1995), no. 1, 229–274.

[12] Peter Buser, *Geometry and spectra of compact Riemann surfaces*, Progress in Mathematics, vol. 106, Birkhäuser Boston Inc., 1992.

[13] John R. Cannon, *The one-dimensional heat equation*, Encyclopedia of Mathematics and its Applications, vol. 23, Addison-Wesley Publishing Company Advanced Book Program, Reading, MA, 1984.

[14] Isaac Chavel, *Eigenvalues in Riemannian geometry*, Pure and Applied Mathematics, vol. 115, Academic Press Inc., Orlando, FL, 1984.

[15] Isaac Chavel, *Riemannian geometry—a modern introduction*, Cambridge Tracts in Mathematics, vol. 108, Cambridge University Press, 1993.

[16] Kevin Corlette, *Hausdorff dimensions of limit sets. I*, Invent. Math. **102** (1990), no. 3, 521–541.

[17] Kevin Corlette, *Archimedean superrigidity and hyperbolic geometry*, Ann. of Math. (2) **135** (1992), no. 1, 165–182.

[18] E. Brian Davies, *Pointwise bounds on the space and time derivatives of heat kernels*, J. Operator Theory **21** (1989), no. 2, 367–378.

[19] E. Brian Davies, *Heat kernels and spectral theory*, Cambridge Tracts in Mathematics, vol. 92, Cambridge University Press, 1990.

[20] E. Brian Davies, *L^p spectral independence and L^1 analyticity*, J. London Math. Soc. (2) **52** (1995), no. 1, 177–184.

[21] E. Brian Davies, *L^p spectral theory of higher-order elliptic differential operators*, Bull. London Math. Soc. **29** (1997), no. 5, 513–546.

[22] E. Brian Davies & Nikolaos Mandouvalos, *Heat kernel bounds on manifolds with cusps*, J. Funct. Anal. **75** (1987), no. 2, 311–322.

[23] E. Brian Davies & Nikolaos Mandouvalos, *Heat kernel bounds on hyperbolic space and Kleinian groups*, Proc. London Math. Soc. (3) **57** (1988), no. 1, 182–208.

[24] E. Brian Davies, Barry Simon, & Michael E. Taylor, *L^p spectral theory of Kleinian groups*, J. Funct. Anal. **78** (1988), no. 1, 116–136.

[25] Andreas Defant & Klaus Floret, *Tensor norms and operator ideals*, North-Holland Mathematics Studies, vol. 176, North-Holland Publishing Co., Amsterdam, 1993.

[26] Jozef Dodziuk, *Maximum principle for parabolic inequalities and the heat flow on open manifolds*, Indiana Univ. Math. J. **32** (1983), no. 5, 703–716.

[27] Harold Donnelly, *Asymptotic expansions for the compact quotients of properly discontinuous group actions*, Illinois J. Math. **23** (1979), no. 3, 485–496.

[28] Harold Donnelly & Peter Li, *Pure point spectrum and negative curvature for noncompact manifolds*, Duke Math. J. **46** (1979), no. 3, 497–503.

[29] Patrick B. Eberlein, *Geometry of nonpositively curved manifolds*, Chicago Lectures in Mathematics, University of Chicago Press, Chicago, IL, 1996.

[30] Patrick B. Eberlein & Barrett O'Neill, *Visibility manifolds*, Pacific J. Math. **46** (1973), 45–109.

[31] Jürgen Elstrodt, *Die Resolvente zum Eigenwertproblem der automorphen Formen in der hyperbolischen Ebene. I*, Math. Ann. **203** (1973), 295–300.

[32] Jürgen Elstrodt, *Die Resolvente zum Eigenwertproblem der automorphen Formen in der hyperbolischen Ebene. II*, Math. Z. **132** (1973), 99–134.

[33] Jürgen Elstrodt, *Die Resolvente zum Eigenwertproblem der automorphen Formen in der hyperbolischen Ebene. III*, Math. Ann. **208** (1974), 99–132.

[34] Joseph Fourier, *Théorie analytique de la chaleur*, Éditions Jacques Gabay, Paris, 1988, Reprint of the 1822 original.

[35] Ramesh Gangolli & Veeravalli S. Varadarajan, *Harmonic analysis of spherical functions on real reductive groups*, Ergebnisse der Mathematik und ihrer Grenzgebiete [Results in Mathematics and Related Areas], vol. 101, Springer-Verlag, Berlin, 1988.

[36] Alexander Grigor'yan, *Gaussian upper bounds for the heat kernel on arbitrary manifolds*, J. Differential Geom. **45** (1997), no. 1, 33–52.

[37] Alexander Grigor'yan, *Estimates of heat kernels on Riemannian manifolds*, Spectral theory and geometry (Edinburgh, 1998), London Math. Soc. Lecture Note Ser., vol. 273, Cambridge Univ. Press, 1999, pp. 140–225.

[38] Mikhail Gromov & Ilya I. Piatetski-Shapiro, *Nonarithmetic groups in Lobachevsky spaces*, Inst. Hautes Études Sci. Publ. Math. (1988), no. 66, 93–103.

[39] Mikhail Gromov & Richard Schoen, *Harmonic maps into singular spaces and p-adic superrigidity for lattices in groups of rank one*, Inst. Hautes Études Sci. Publ. Math. (1992), no. 76, 165–246.

[40] Toshiaki Hattori, *Asymptotic geometry of arithmetic quotients of symmetric spaces*, Math. Z. **222** (1996), no. 2, 247–277.

[41] Sigurdur Helgason, *Differential geometry, Lie groups, and symmetric spaces*, Pure and Applied Mathematics, vol. 80, Academic Press Inc., New York, 1978.

[42] Sigurdur Helgason, *Geometric analysis on symmetric spaces*, Mathematical Surveys and Monographs, vol. 39, American Mathematical Society, Providence, RI, 1994.

[43] Rainer Hempel & Jürgen Voigt, *The spectrum of a Schrödinger operator in $L_p(\mathbf{R}^\nu)$ is p-independent*, Comm. Math. Phys. **104** (1986), no. 2, 243–250.

[44] Rainer Hempel & Jürgen Voigt, *On the L_p-spectrum of Schrödinger operators*, J. Math. Anal. Appl. **121** (1987), no. 1, 138–159.

[45] Harro Heuser, *Funktionalanalysis*, third ed., Mathematische Leitfäden., B. G. Teubner, Stuttgart, 1992.

[46] Matthias Hieber, *Gaussian estimates and invariance of the L^p-spectrum for elliptic operators of higher order*, Rend. Istit. Mat. Univ. Trieste **28** (1996), no. suppl., 235–249 (1997).

[47] Hans-Christoph Im Hof, *Die Geometrie der Weylkammern in symmetrischen Räumen vom nichtkompakten Typ*, Habilitationsschrift, Bonn, 1979.

[48] Lizhen Ji & Robert MacPherson, *Geometry of compactifications of locally symmetric spaces*, Ann. Inst. Fourier (Grenoble) **52** (2002), no. 2, 457–559.

[49] Anthony W. Knapp, *Lie groups beyond an introduction*, Progress in Mathematics, vol. 140, Birkhäuser Boston Inc., 1996.

[50] Gerhard Knieper, *On the asymptotic geometry of nonpositively curved manifolds*, Geom. Funct. Anal. **7** (1997), no. 4, 755–782.

[51] Shoshichi Kobayashi & Katsumi Nomizu, *Foundations of differential geometry. Vol. I*, Wiley Classics Library, John Wiley & Sons Inc., New York, 1996, Reprint of the 1963 original, A Wiley-Interscience Publication.

[52] Peer C. Kunstmann, *Heat kernel estimates and L^p spectral independence of elliptic operators*, Bull. London Math. Soc. **31** (1999), no. 3, 345–353.

[53] Peer C. Kunstmann, *Uniformly elliptic operators with maximal L_p-spectrum in planar domains*, Arch. Math. (Basel) **76** (2001), no. 5, 377–384.

[54] Peer C. Kunstmann, *L_p-spectral properties of the Neumann Laplacian on horns, comets and stars*, Math. Z. **242** (2002), no. 1, 183–201.

[55] Robert P. Langlands, *On the functional equations satisfied by Eisenstein series*, Springer-Verlag, Berlin, 1976, Lecture Notes in Mathematics, Vol. 544.

[56] Enrico Leuzinger, *Kazhdan's property (T), L^2-spectrum and isoperimetric inequalities for locally symmetric spaces*, Comment. Math. Helv. **78** (2003), no. 1, 116–133.

[57] Enrico Leuzinger, *Critical exponents of discrete groups and L^2-spectrum*, Proc. Amer. Math. Soc. **132** (2004), no. 3, 919–927 (electronic).

[58] Enrico Leuzinger, *Tits geometry, arithmetic groups, and the proof of a conjecture of Siegel*, J. Lie Theory **14** (2004), no. 2, 317–338.

[59] Vitali A. Liskevich & M. A. Perel'muter, *Analyticity of sub-Markovian semigroups*, Proc. Amer. Math. Soc. **123** (1995), no. 4, 1097–1104.

[60] Gregory A. Margulis, *Discrete subgroups of semisimple Lie groups*, Ergebnisse der Mathematik und ihrer Grenzgebiete (3) [Results in Mathematics and Related Areas (3)], vol. 17, Springer-Verlag, Berlin, 1991.

[61] Werner Müller, *Manifolds with cusps of rank one*, Lecture Notes in Mathematics, vol. 1244, Springer-Verlag, Berlin, 1987, Spectral theory and L^2-index theorem.

[62] Peter J. Nicholls, *The ergodic theory of discrete groups*, London Mathematical Society Lecture Note Series, vol. 143, Cambridge University Press, 1989.

[63] Barrett O'Neill, *Semi-Riemannian geometry*, Pure and Applied Mathematics, vol. 103, Academic Press Inc., New York, 1983, With applications to relativity.

[64] Patrick Ostellari, *Estimations globales du noyau de la chaleur*, Ph.D. thesis, Université Henri Poincaré, Nancy 1, 2003.

[65] Samuel J. Patterson, *The limit set of a Fuchsian group*, Acta Math. **136** (1976), no. 3-4, 241–273.

[66] Peter Petersen, *Riemannian geometry*, Graduate Texts in Mathematics, vol. 171, Springer-Verlag, New York, 1998.

[67] John G. Ratcliffe, *Foundations of hyperbolic manifolds*, Graduate Texts in Mathematics, vol. 149, Springer-Verlag, New York, 1994.

[68] Michael Reed & Barry Simon, *Tensor products of closed operators on Banach spaces*, J. Functional Analysis **13** (1973), 107–124.

[69] Steven Rosenberg, *The Laplacian on a Riemannian manifold*, London Mathematical Society Student Texts, vol. 31, Cambridge University Press, 1997.

[70] Takashi Sakai, *Riemannian geometry*, Translations of Mathematical Monographs, vol. 149, American Mathematical Society, Providence, RI, 1996.

[71] Robert S. Strichartz, *Analysis of the Laplacian on the complete Riemannian manifold*, J. Funct. Anal. **52** (1983), no. 1, 48–79.

[72] Karl-Theodor Sturm, *On the L^p-spectrum of uniformly elliptic operators on Riemannian manifolds*, J. Funct. Anal. **118** (1993), no. 2, 442–453.

[73] Dennis Sullivan, *The density at infinity of a discrete group of hyperbolic motions*, Inst. Hautes Études Sci. Publ. Math. (1979), no. 50, 171–202.

[74] Dennis Sullivan, *Related aspects of positivity in Riemannian geometry*, J. Differential Geom. **25** (1987), no. 3, 327–351.

[75] Michael E. Taylor, *L^p-estimates on functions of the Laplace operator*, Duke Math. J. **58** (1989), no. 3, 773–793.

[76] Veeravalli S. Varadarajan, *Harmonic analysis on real reductive groups*, Springer-Verlag, Berlin, 1977, Lecture Notes in Mathematics, Vol. 576.

[77] Nicholas Th. Varopoulos, *Analysis on Lie groups*, J. Funct. Anal. **76** (1988), no. 2, 346–410.

[78] Jürgen Voigt, *The sector of holomorphy for symmetric sub-Markovian semigroups*, Functional analysis (Trier, 1994), de Gruyter, Berlin, 1996, pp. 449–453.

[79] Dave Witte Morris, *Introduction to Arithmetic Groups*, URL-Address: http://www.math.okstate.edu/~dwitte, February 2003.

[80] Joseph A. Wolf, *Spaces of constant curvature*, McGraw-Hill Book Co., New York, 1967.

[81] Robert J. Zimmer, *Ergodic theory and semisimple groups*, Monographs in Mathematics, vol. 81, Birkhäuser Verlag, Basel, 1984.

Printed in Dunstable, United Kingdom

91343139R00063